Temperature Calculation in Fire Safety Engineering

Ulf Wickström

Temperature Calculation in Fire Safety Engineering

Ulf Wickström
Luleå University of Technology
Luleå, Sweden

Instructor answers to review questions can be found at
http://www.springer.com/us/book/9783319301709

ISBN 978-3-319-80738-6 ISBN 978-3-319-30172-3 (eBook)
DOI 10.1007/978-3-319-30172-3

Softcover reprint of the hardcover 1st edition 2016

Printed on acid-free paper

This Springer imprint is published by Springer Nature
The registered company is Springer International Publishing AG Switzerland

Preface

This book is about temperature calculation and heat transfer. It is intended for researchers, students and teachers in the field of fire safety engineering as well as consultants and others interested in analysing and understanding fire and temperature developments. It gives a consistent scientific background to engineering calculation methods applicable to analyses for both materials' reaction to fire and fire resistance of structures. Several new formulas and diagrams facilitating calculations are presented.

The book is particularly devoted to problems involving severe thermal conditions as are of interest in fire dynamics and FSE. However, definitions, nomenclature and theories used are aligned with those of general textbooks on temperature calculation and heat transfer such as [1, 2].

In particular great effort has been put on defining boundary conditions in a correct and suitable way for calculations. A large portion of the book is devoted to boundary conditions and measurements of thermal exposure by radiation and convection. Thus, the concept and theory of adiabatic surface temperature and measurements of temperature with plate thermometers are thoroughly explained.

Initially a number of zero- and one-dimensional cases assuming constant material properties are dealt with where exact closed form analytical solutions are possible. These can, however, generally only be used for estimates in FSE problems as they require assumptions of constant material properties and boundary conditions. In most cases numerical calculations are therefore needed for considering material properties changing with temperature and non-linear boundary conditions due to emission of radiant heat. Thus, several recursion formulas are given in the book which are suited for spreadsheet calculation codes (such as MS Excel). For more advanced calculations, introductions and guidance are given to finite element analyses.

The phenomena of heat transfer by radiation and convection are introduced based on what can be found in general textbooks. Several of the formulas are, however, adapted to FSE problems, and unique charts and tables are presented which considerably facilitates calculations.

A renewed method for modelling compartment fires is presented which has led to simple and accurate prediction tools for both pre- and post-flashover fires.

The final three chapters deal with temperature calculations in steel, concrete and timber structures exposed to standard time-temperature fire curves. Handy temperature calculation tools are presented, and several examples are shown on how the finite element code TASEF can be used to calculate temperature in various configurations.

Luleå, Sweden Ulf Wickström

Contents

Nomenclature[1]

a	Heat of vaporization [J/kg]
A	Area [m^2]
Bi	Biot number[−]
c	Specific heat (capacity) [W s/(kg K)] or [J/(kg K)]
C	Heat capacity per unit area [W s/(m^2 K)]
d	Thickness [m]
D	Diameter [m]
f	Forcing function
F	View factor [−]
Gr	Grashof number [−]
h	Heat transfer coefficient [W/(m^2 K)]
k	Conductivity [W/(m K)]
l	Latent heat [W s/m^3]
L	Length [m]
Nu	Nusselt number [−]
O	Opening factor [$m^{1/2}$]
P	Perimeter of surface [m]
Pr	Prandtl number [−]
q	Heat [W s or J]
R	One-dimensional thermal resistance [(K m^2)/W]
Re	Reynolds number [−]
T	Temperature [K] or [°C]
V	Volume [m^3]
x	Length [m]
α	Thermal diffusivity [m^2/s]
α	Absorption coefficient [−]
Γ	Compartment fire time factor [−]

[1] Definitions of symbols are given throughout the text. Some selected symbols are listed below.

δ	Boundary layer thickness [m]
ε	Emissivity [−]
η	Reduction coefficient [−]
κ	Absorption coefficient [m^{-1}]
ν	Kinetic viscosity [m^2/s]
ρ	Density [kg/m^3]
σ	Stefan-Boltzmann constant ($5.67 \cdot 10^{-8}$) [$W/(m^2 K^4)$]
τ	Time constant [s]
χ	Combustion efficiency [−]

Superscripts

$'$	Per unit length
$''$	Per unit area
$'''$	Per unit volume
$\cdot$	Per unit volume, time derivative
$-$	Vector, matrix

Subscripts

0	Surface ($x = 0$)
∞	Ambient
AST	Adiabatic surface temperature
B	Burning
CC	Cone calorimeter
Con, c	Convection
cr	Critical
d	Duration
emi	Emitted
hfm	Heat flux meter
f	Film
f	Fire
gas	Gas
i	Initial
ig	Ignition
in	Insulation
inc	Incident (radiation)
L	Air, gas convection
o	Opening
p	Constant pressure
p	Plume
PT	Plate thermometer
r	Radiation

rad	Net radiation
RC	Room/corner test
s	Surface
sh	Shield
st	Steel
TC	Thermocouple
tot	Total, radiation + convection
ult	Ultimate
w	Wall, surrounding boundary

Abbreviations

AST	Adiabatic surface temperature
ASTM	ASTM International, earlier *American Society for Testing and Materials*
CEN, EN	European Committee for Standardization developing EN standards
EUROCODE	EN Eurocodes is a series of 10 European Standards for the design of buildings
FSE	Fire safety engineering
ISO	International Organization for Standardization
PT	Plate thermometer
TASEF	Computer code for Temperature Analysis of Structures Exposed to Fire

Chapter 1
Introduction

Temperature is the dominating factor in determining the rate and extent of chemical reactions including breakdown of organic compounds and deteriorations of strength and stiffness of structural materials such as steel and concrete. Phase change phenomena including ignition as well as severe loss of strength of materials are often related to specific elevated temperature levels. Temperatures of fire gases are also of crucial importance as they initiate gas movements thereby spread of smoke and toxic fire gases. Fire temperatures vary typically over several hundred degrees. Therefore a number of thermal phenomena need special attention such as phase changes of materials and heat transfer by radiation when calculating temperature of fire-exposed materials.

In this chapter some of the basic concepts of heat transfer are briefly introduced. More detailed presentations are given in following chapters. A summary of the principles of electric circuit analogy which is used throughout this book is also given as well as some general comments on material properties.

1.1 Basic Concepts of Temperature, Heat and Heat Flux

Temperature is an *intensive* or bulk property, i.e. a physical property that does not depend on size or the amount of material in a system. It is scale invariant. By contrast, heat is an *extensive* property which is directly proportional to the amount of material in a system. Density is another example of an intensive quantity as it does not depend on the quantity, while mass and volume are extensive quantities.

In the presentation below the thermal material properties c, *specific heat capacity* or just *specific heat*, and ρ, *density*, are assumed constant.

U. Wickström, *Temperature Calculation in Fire Safety Engineering*,
DOI 10.1007/978-3-319-30172-3_1

1.1.1 Heat and Temperature

The heat q of a body is proportional to the mass and the temperature rise T.

$$q = c \cdot m \cdot T \tag{1.1}$$

and with the volume V and the material density ρ

$$m = \rho \cdot V \tag{1.2}$$

the heat of a body may be written as

$$q = c \cdot \rho \cdot V \cdot T \tag{1.3}$$

In a more general way where c and ρ may vary with temperature and location, the heat of a body may be written as an integral over temperature range and volume as

$$q = \int_T \left(\int_V c \cdot \rho \cdot dV \right) dT \tag{1.4}$$

1.1.2 Heat Transfer Modes

Heat is transferred in three modes, *conduction*, *convection* and *radiation*. The concept of thermal conduction can be seen as a molecular process by which energy is transferred from particles of high energy/temperature to particles of low energy/temperature. High temperatures are associated with higher molecular energies, and when neighbouring molecules collide a transfer of energy from the more to the less energetic molecules occurs. This process takes place in fluids as well although the main mode of heat transfer then is generally due to motion of matter, i.e. convection.

By the definition of temperature, heat is transferred from places with higher temperatures to places with lower temperature, i.e. the temperature difference is the driving force of the heat transfer. In one dimension the heat flux $\dot{q}''$ across a plane wall with the thickness L and a conductivity k may be written as

$$\dot{q}'' = \frac{k}{L} \cdot (T_1 - T_2) \tag{1.5}$$

Notice that the superscript ($''$) denotes per unit area and the accent character ($\dot{}$) per unit time. Under steady-state conditions the temperature distribution will be linear as shown in Fig. 1.1.

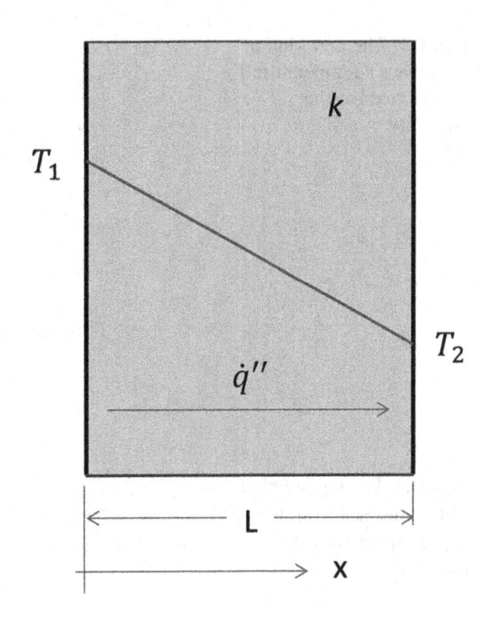

Fig. 1.1 Steady-state temperature distribution and heat flux across a plane wall according to Eqs. 1.5 and 1.6

In differential form Eq. 1.5 may be written as

$$\dot{q}_x'' = -k \cdot \frac{dT}{dx} \tag{1.6}$$

This is the *Fourier's law of heat conduction* which implies that the heat flux is proportional to the heat conductivity of the material and the thermal gradient.

1.1.3 The Three Kinds of Boundary Conditions

In addition to the differential equation valid for the interior, boundary conditions must be specified when calculating temperatures in solids. A thorough understanding of how to express BCs is particularly important in FSE.

In principle there are three kinds of BCs denoted first, second and third [2]. The first kind is prescribed temperature, the second kind is prescribed heat flux and the third kind is heat flux dependent on the difference between prescribed surrounding gas or fluid temperatures and the current boundary or surface temperature. The latter type of BC is by far the most common in FSE. It may include heat transfer by convection as well as radiation. More on boundary conditions relevant in FSE, see Chap. 4 and for more details on radiation and convection, see Chaps. 5 and 6, respectively.

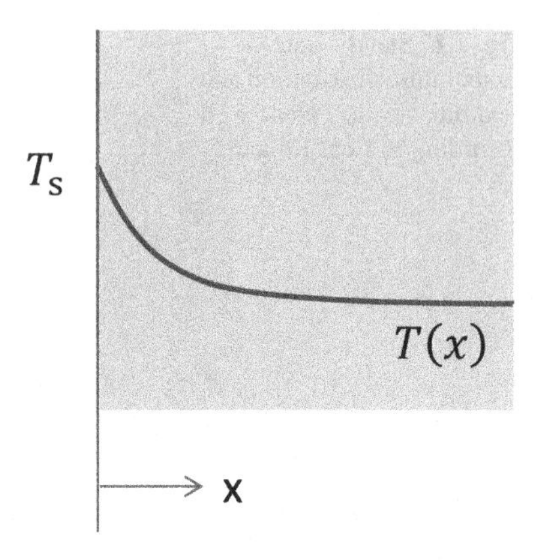

Fig. 1.2 The *first* kind of BC means a temperature T_s is prescribed at the boundary

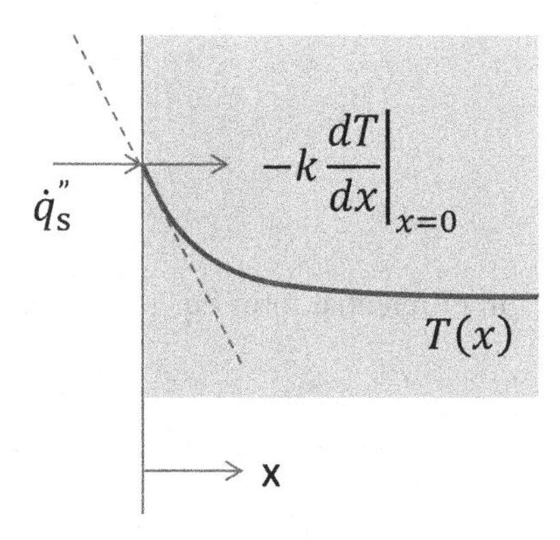

Fig. 1.3 The *second* kind of BC means a heat flux $\dot{q}_s''$ is prescribed at the boundary

The *first* kind of BC (sometimes called the Dirichlet boundary condition) as shown in Fig. 1.2 means a temperature T_s is prescribed at the boundary ($x = x_0$), i.e.

$$T(x_0) = T_s \tag{1.7}$$

In fire engineering it may, for example, be assumed when a surface of a light insulating material is exposed to fire. The surface temperature may then be approximated to adjust momentarily to the boundary gas and radiation temperatures which facilitates the computations.

The *second* kind of BC (sometimes called the Neumann boundary condition) as shown in Fig. 1.3 means a heat flux $\dot{q}_s''$ is prescribed at the boundary, i.e.

$$\dot{q}_s'' = -k \cdot \frac{\partial T}{\partial x}\bigg|_{x=x_0} \tag{1.8}$$

Thus the heat flux to the boundary is equal the heat being conducted away from the surface into the solid according to the Fourier's law, or in the case of lumped heat, it is approximated as the heat stored, see Chap. 3. A special case of the second

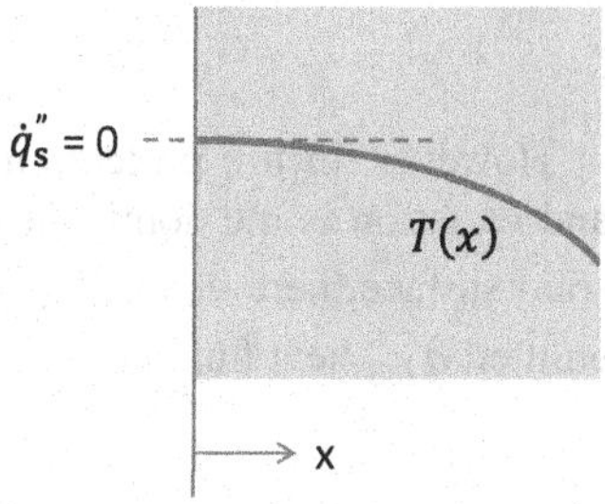

Fig. 1.4 An adiabatic surface, i.e. a perfectly insulating surface, or a surface along a line of symmetry is a special case of a second kind of BC

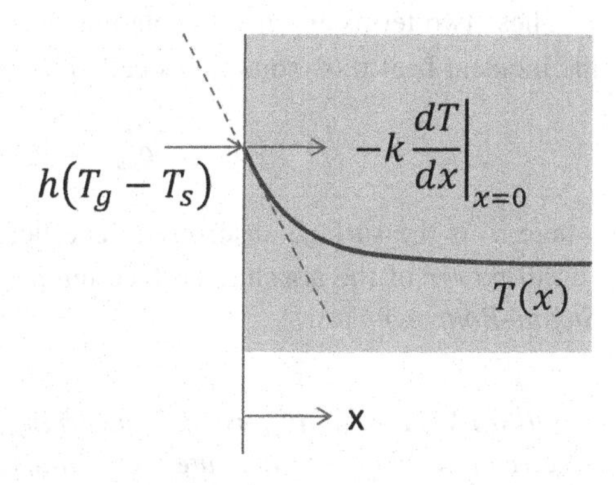

Fig. 1.5 The *third* kind of BC means the heat flux to the boundary depends on the difference between prescribed surrounding gas or fluid temperatures and current surface temperature

kind of BC is an *adiabatic surface* or a perfectly insulated surface, or a surface along a line of symmetry where the heat flux by definition of symmetry is zero, see Fig. 1.4.

In FSE the second kind of BC is rarely applicable. The concept of "heat flux" meant as heat flux to a surface kept at ambient is, however, often used as a general measure of thermal exposure. This is in reality a third kind of BC but unfortunately it is difficult to apply as a boundary condition for temperature calculations, see Sect. 9.2.1.

The *third* kind of BC (sometimes called the Robin boundary condition) means the heat flux to the boundary surface depends on the difference between prescribed surrounding gas or fluid temperatures and the current boundary or surface temperature, see Fig. 1.5. It is sometimes also called *natural boundary conditions* or *Newton's law of cooling*. In the simplest form the heat transfer is proportional to the difference between the surrounding gas temperature and the surface temperature. The proportionality constant h is denoted the *heat transfer coefficient*.

$$\dot{q}'' = -k \cdot \frac{\partial T}{\partial x}\Big|_{x=x_0} = h(T_g - T_s) \quad (1.9)$$

Equation 1.9 is a reasonable approximation when heat transfer by convection only is considered. Therefore we write the heat flux by convection $\dot{q}''_{con}$ as

$$\dot{q}''_{con} = h(T_g - T_s) \tag{1.10}$$

However, in fire protection engineering problems temperature is usually high and radiation is the dominant mode of heat transfer. The net heat flux entering a solid surface, here denoted $\dot{q}''_{rad}$, is the difference between the absorbed $\dot{q}''_{abs}$ and emitted $\dot{q}''_{emi}$ heat flux, i.e.

$$\dot{q}''_{rad} = \dot{q}''_{abs} - \dot{q}''_{emi} \tag{1.11}$$

These two terms are in principle independent. The absorbed flux is a portion of the incident heat flux (sometimes called *irradiance*) $\dot{q}''_{inc}$ to a surface. Thus

$$\dot{q}''_{abs} = \alpha_s \cdot \dot{q}''_{inc} \tag{1.12}$$

where α_s is the surface absorptivity coefficient. The emitted heat depends on the fourth power of the absolute surface temperature T_s^4 (in Kelvin) according to the *Stefan–Boltzmann law*:

OBSERVE that in all formula concerning radiation the temperature must be given in Kelvin [*K*], *absolute temperature.*

$$\dot{q}''_{emi} = \varepsilon_s \cdot \sigma \cdot T_s^4 \tag{1.13}$$

where ε_s is the surface emissivity, and the physical constant $\sigma = 5.67 \cdot 10^{-8}$ W/(m^2 K^4) is named the *Stefan–Boltzmann constant.*

The surface properties ε_s and α_s have values between 0 and 1 and are according to *Kirchhoff's identity* equal, i.e.

$$\alpha_s = \varepsilon_s \tag{1.14}$$

(The Kirchhoff's identity does not apply when the source emitting radiation to a surface and the target surface have very different temperatures. Then Eqs. 1.12 and 1.13 must be used with α and ε depending on the wavelength spectrums of the incident radiation and the emitted radiation, respectively. This condition is rarely considered in FSE as the absorptivity/emissivity of most building materials changes only marginally with the temperature, the radiation is considered to be gray). Thus by inserting Eqs. 1.12 and 1.13 into Eq. 1.11, the *net radiation heat flux* to a surface can be written as

$$\dot{q}''_{rad} = \varepsilon_s \cdot \left(\dot{q}''_{inc} - \sigma \cdot T_s^4\right) \quad (1.15)$$

Alternatively, the neat radiation heat flux may also be expressed in terms of as

$$\dot{q}''_{rad} = \varepsilon_s \cdot \sigma\left(T_r^4 - T_s^4\right) \quad (1.16)$$

where T_r is the *incident black body radiation temperature* or just the *black body radiation temperature* defined by the identity

$$\dot{q}''_{inc} \equiv \sigma \cdot T_r^4 \quad (1.17)$$

The heat flux by radiation and convection can be superimposed to form the total heat flux which in this book is denoted $\dot{q}''_{tot}$. Then BC the third kind becomes

$$\dot{q}''_{tot} = \dot{q}''_{rad} + \dot{q}''_{con} \quad (1.18)$$

and thus

$$\dot{q}''_{tot} = \varepsilon_s \cdot \left(\dot{q}''_{inc} - \sigma \cdot T_s^4\right) + h\left(T_g - T_s\right) \quad (1.19)$$

or alternatively

$$\dot{q}''_{tot} = \varepsilon_s \cdot \sigma \cdot \left(T_r^4 - T_s^4\right) + h\left(T_g - T_s\right) \quad (1.20)$$

Equation 1.20 is a *mixed boundary condition* as it contains independent heat transfer by radiation and convection. In standards on fire resistance of structures such as Eurocode 1, EN 1991-1-2, the radiation and gas fire temperatures are assumed equal, T_f, and then Eq. 1.20 becomes

$$\dot{q}''_{tot} = \varepsilon_s \cdot \sigma \cdot \left(T_f^4 - T_s^4\right) + h\left(T_f - T_s\right) \quad (1.21)$$

Notice that, as the heat emitted from a surface $\left(\varepsilon_s \cdot \sigma \cdot T_s^4\right)$ depends on the forth power of the surface temperature, the problems become mathematically non-linear which prohibits exact analytical solutions of the heat equation. To avoid this, the introductory Chaps. 2 and 3 are limited to cases where the heat transfer coefficient h can be assumed constant.

More on boundary conditions are given in Chap. 4, and methods for calculating heat transfer by radiation and convection are given in Chaps. 5 and 6.

1.1.4 Transient or Unsteady-State Heat Conduction

Heat is transferred by conduction, convection or radiation at heat flow rates denoted $\dot{q}\cdot$. Then the changes of heat content dq of a body over a time interval dt becomes

$$dq = \dot{q} \cdot dt \tag{1.22}$$

By differentiating Eq. 1.3

$$\dot{q} = c \cdot \rho \cdot V \cdot \frac{dT}{dt} \tag{1.23}$$

Figure 1.6 shows a one-dimensional increment dx. The heat entering from the left side is $\dot{q}''_x$ and the heat leaving on the right side $\dot{q}''_{x+dx}$. Hence

$$\dot{q}'' = -k \cdot \frac{dT}{dx} + \left[k \cdot \frac{dT}{dx}, + \frac{d}{dx}\left(k \cdot \frac{dT}{dx}\right)\right] \cdot dx = \frac{d}{dx}\left(k \cdot \frac{dT}{dx}\right) \cdot dx \tag{1.24}$$

The difference is the change of heat being stored per unit time, i.e.

$$\dot{q}'' = c \cdot \rho \cdot \frac{dT}{dt} \cdot dx \tag{1.25}$$

Now by combining Eqs. 1.24 and 1.25, the *heat conduction equation* or the *heat diffusion equation* in one dimension is obtained as

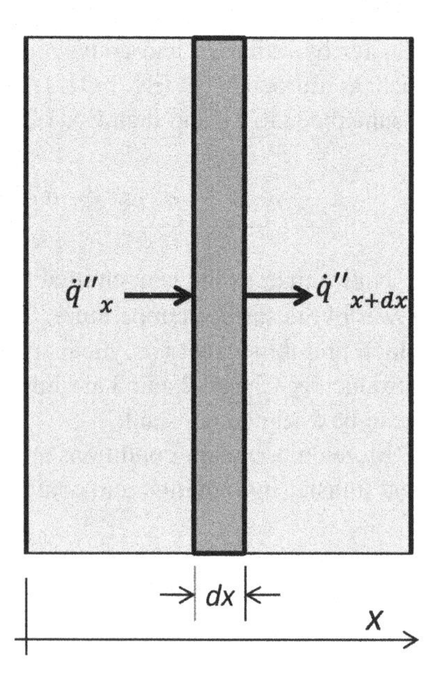

Fig. 1.6 One-dimensional increment

$$\frac{d}{dx}\left(k \cdot \frac{dT}{dx}\right) = c \cdot \rho \cdot \frac{dT}{dt} \tag{1.26}$$

or if k is constant, the heat conduction equation may be written as

$$\frac{d^2T}{dx^2} = \frac{1}{\alpha} \cdot \frac{dT}{dt} \tag{1.27}$$

where α is the *thermal diffusivity*, a parameter group defined as

$$\alpha = \frac{k}{c \cdot \rho} \tag{1.28}$$

with the dimension m^2/s in SI units.

In three dimensions *x, y and z* the general heat conduction equation is

$$\frac{\partial}{\partial x}\left(k \cdot \frac{dT}{dx}\right) + \frac{\partial}{\partial y}\left(k \cdot \frac{dT}{dy}\right) + \frac{\partial}{\partial z}\left(k \cdot \frac{dT}{dz}\right) = c \cdot \rho \cdot \frac{dT}{dt} \tag{1.29}$$

The heat capacity c and the density ρ appear always as a product in heat diffusion equations, sometimes denoted *specific volumetric heat capacity* ($J/(m^3$ K) in SI units). Alternatively Eq. 1.26 may be written as

$$\frac{d}{dx}\left(k \cdot \frac{dT}{dx}\right) = \frac{de}{dt} \tag{1.30}$$

where e is the heat content per unit volume named the *specific volumetric enthalpy*. By definition it is the heat needed to rise the temperature of a unit volume from one level (e.g. 0 °C) to a higher temperature. Then

$$e = \int_0^T c \cdot \rho \cdot dT \tag{1.31}$$

For materials with $c \cdot \rho$ constant and independent of temperature the volumetric specific enthalpy becomes

$$e = c \cdot \rho \cdot T \tag{1.32}$$

The concept of specific volumetric enthalpy is advantageous to use when considering physical and chemical transformations. Then numerical temperature calculations may be facilitated as will be presented in more detail in Chap. 7.

In the simplest cases of transient heat transfer problems, the temperature in a body is assumed uniform. Then as only a single uniform temperature is calculated with no variation depending on position, this type of problems is *zero-dimensional*. More on lumped heat calculations can be found in Sects. 3.1, 7.1 and 13.3.

1.2 Electric Circuit Analogy in One Dimension

There are analogies between parameters of heat transfer systems and electric circuits. These will be used throughout this book to illustrate, develop and explain various temperature and heat transfer calculation formulas. An overview of corresponding parameters, nomenclature and icons of resistance and capacitance is given in Table 1.1. Notice that the resistance R for thermal problems refers to a unit area while the analogue electric resistance includes the area in R_e. In summary, temperature is analogue to electric potential or voltage, heat flow to electric current, thermal resistance to electric resistance and heat capacity to electric capacity.

From the discipline of electric circuits the rules of combining resistances can be applied. Thus two resistances *in series* between A and C as shown in Fig. 1.7 can be summarized as

Table 1.1 Analogies between thermal parameters in one dimension and electric parameters and units

Temperature and heat		Electric circuit analogy	
Parameter and nomenclature	SI units	Parameter and nomenclature	SI units
Heat, q	[J or Ws]	Electric charge, Q	[J]
Temperature, T	[K or °C]	Electric potential, U	[V]
Heat flow, $\dot{q}$	[W]	Electric current, I	[A]
$\dot{q} = A \cdot \frac{\Delta T}{R}$		$I = \frac{\Delta U}{R_e}$ (Ohm's law)	
T_1 R T_2		U_1 R_e U_2	
Thermal resistance		Resistor	
$\Delta T = (T_1 - T_2)$		$\Delta U = (U_1 - U_2)$	
Heat flux, $\dot{q}''$	[W/m^2]	Electric current per unit area, I/A	[A/m^2]
$\dot{q}'' = \frac{\Delta T}{R}$			
1D thermal resistance, R	[m^2 K/W]	Electric resistance, R_e	[Ω]
Surface resistance $R_h = \frac{1}{h}$	[m^2 K/W]		
Solid resistance $R_k = \frac{d}{k}$	[m^2 K/W]		
1D heat capacitance, C $C = c \cdot \rho \cdot d$ $q'' = C \cdot T$ T C T=0 Lumped-heat-capacity	[J/(m^2 K)] [J/(m^2 K)] [J/m^2]	Electric capacitance, C_e $Q = C_e \cdot U$ U C_e Earth U=0 Electric capacitor	[J/V] [J]

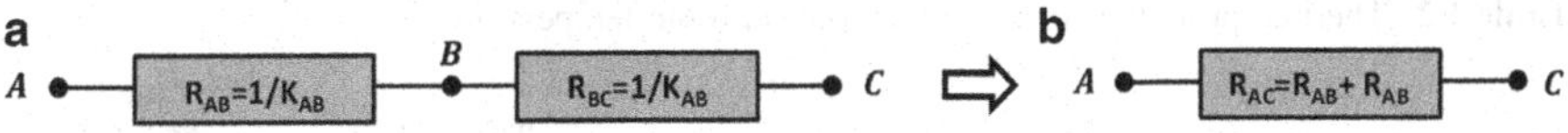

Fig. 1.7 Rules for combining resistances in series according to Eq. 1.33. (a) Resistances in series. (b) Resultant resistance

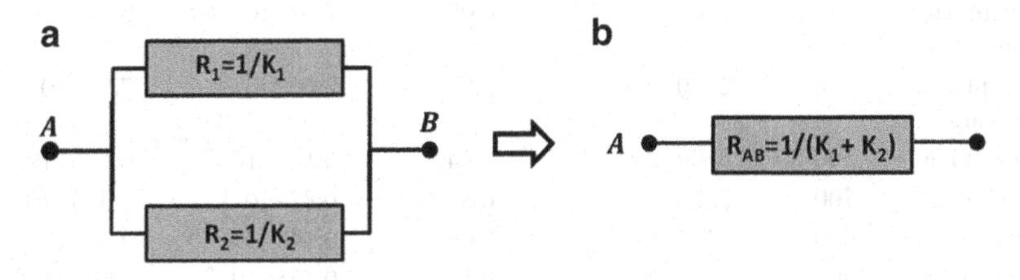

Fig. 1.8 Rules for combining parallel resistances according to Eq. 1.34. (a) Parallel resistances. (b) Resultant resistance

$$R_{AC} = R_{AB} + R_{BC} = \frac{1}{K_{AB}} + \frac{1}{K_{BC}} \tag{1.33}$$

K denotes the reciprocal of the thermal resistance which could be a heat transfer coefficient, h, or conductivity over a thickness, k/d.

In the case of *parallel resistances* as shown in Fig. 1.8, the resultant resistance between A and B becomes

$$R_{AB} = \frac{1}{\frac{1}{R_1} + \frac{1}{R_2}} = \frac{1}{K_1 + K_2} \tag{1.34}$$

Thus as an example according to Fig. 1.8, the heat flux between A and B may be written as

$$\dot{q}'' = (K_1 + K_2) \cdot (T_A - T_B) = \frac{1}{\frac{1}{R_1} + \frac{1}{R_2}} \cdot (T_A - T_B) \tag{1.35}$$

where T_A and T_B are the temperatures at point A and B, respectively. K_1 and K_2 may, for example, be heat transfer coefficients due to radiation and convection.

1.3 Material Properties at Elevated Temperature

The flow of heat by conduction in a body is proportional to the thermal conductivity of the material and the temperature gradient according to Fourier's law as given by Eq. 1.6. Under steady-state conditions the conductivity denoted k is the sole material property while under transient conditions the density ρ and the specific

Table 1.2 Thermal properties of some materials at room temperature

Material	Density ρ [kg/m^3]	Specific heat capacity c [J/(kg K)]	Conductivity k [W/(m K)]	Thermal diffusivity $k/(\rho \cdot c)$ [m^2/s]	Thermal inertia $k \cdot \rho \cdot c$ [(W^2 s)/(m^4 K^2)]
Air	1.23	1010	0.024	$19.3 \cdot 10^{-6}$	$0.030 \cdot 10^3$
Polyurethane foam	20	1400	0.03	$1.07 \cdot 10^{-6}$	$0.840 \cdot 10^3$
Fibre insulating board	100	2000	0.04	$2.00 \cdot 10^{-6}$	$7.92 \cdot 10^3$
Wood, pine	500	2800	0.14	$0.100 \cdot 10^{-6}$	$0.196 \cdot 10^6$
Wood, oak	700	2800	0.17	$0.87 \cdot 10^{-6}$	$0.336 \cdot 10^6$
Water	1000	4181	0.604	$0.144 \cdot 10^{-6}$	$2.53 \cdot 10^6$
Gypsum plaster	1400	840	0.5	$0.425 \cdot 10^{-6}$	$0.593 \cdot 10^6$
Concrete	2300	900	1.7	$0.82 \cdot 10^{-6}$	$3.53 \cdot 10^6$
Aluminium	2700	900	200	$82.3 \cdot 10^{-6}$	$486 \cdot 10^6$
Steel (mild)	7850	460	46	$12.7 \cdot 10^{-6}$	$166 \cdot 10^6$
Copper	8930	390	390	$112 \cdot 10^{-6}$	$1362 \cdot 10^6$

Values of this table are only indicative and not necessarily recommended for use in real FSE applications

heat capacity c are needed in addition. In general the thermal conductivity of a solid is bigger than that of a liquid, which is larger than that of a gas. Materials with a low density have in general low heat conductivity while materials with high densities and in particular metals have high thermal conductivities. Insulating materials have low densities and are by definition pure conductors of heat. Table 1.2 shows in the order of density the thermal properties of a number of materials. With the exception of metals, air and water we can derive from this table the very approximate relations between the density ρ and the conductivity k as

$$k \approx 0.04 \cdot e^{0.0017 \cdot \rho} \tag{1.36}$$

The *specific heat* c of a material or substance is the amount of heat needed to change the temperature of a unit mass of the substance by 1°. It is an intensive parameter with the unit of energy per unit mass and degree, in SI units [J/(kg K)] or [Ws/(kg K)]. (This is unlike the extensive variable *heat capacity* (denoted C), which depends on the quantity of material and is expressed in [J/K]). As a general rule the specific heat decreases with density, i.e. it is high for low density materials and low for high density materials. Cementitious materials have a specific heat capacity c slightly under 1000 J/(kg K) while the corresponding values for wood are considerably higher. For metals c is significantly lower and varies inversely with the density. Notice in Table 1.2 that c of water is relatively high, more than four times higher than c for concrete. Therefore the moisture content of a material has great influence on the temperature development. The major influence of the water is, however, when it vaporizes at temperatures exceeding 100 °C.

The product $k \cdot \rho \cdot c$, denoted *thermal inertia*, see Sect. 8.2.2, has a great impact on ignition and flame spread propensities of materials. When the density ρ increases so does normally the conductivity k as well, and consequently the thermal inertia is greatly dependent on the density. It varies over a wide range and therefore the density is a very significant indicator of the fire properties materials.

Notice that the thermal inertia, $k \cdot \rho \cdot c$, of wood is in the order of 300 times as high as the corresponding value of an efficient insulating material such as polyurethane foam. This difference will give these materials a considerable difference in their ignition properties as will be discussed in Chap. 8.

The data given in Table 1.2 refer to room temperature. At elevated temperatures which are relevant in fires and fire-exposed structures the material properties may vary significantly. In addition the parameter values listed cannot be assumed to fully reflect the properties of all materials within any generic class. Specific data for particular products may be provided by the manufacturers.

1.3.1 Structural Materials

The temperature of structures exposed to fully developed fires with gas temperatures reaching 800–1200 °C will gradually increase and eventually the structures may lose their load-bearing capacity as well as their ability to keep fires within confined spaces. In building codes fire resistance requirements are usually expressed in terms of the time a structural element can resist a nominal or standard fire as defined, e.g. in the international standard ISO 834 or the corresponding European standard EN 1363-1. In the USA and Canada the corresponding standard curve for determining fire resistance of building components is given in ASTM E-119. The standard time–temperature curves as defined by the ISO/EN and ASTM standards are shown in Fig. 1.9. More on standard fires can be found in Chap. 12.

Below some general remarks are given for the most common structural materials. Methods for calculating temperature steel, concrete and timber structures exposed to fire are outlined in Chaps. 13, 14 and 15, respectively.

Steel starts to lose both strength and stiffness at about 400 °C and above 600 °C more than half of its original strength is lost, see, e.g. Eurocode 3 [3] or the SFPE Handbook on Fire Protection Engineering [4, 5]. Therefore structural steel elements must in most cases be fire protected by sprayed on compounds, boards, mineral wool or intumescent paint to keep sufficient load-bearing capacity over time when exposed to fire. An example of a steel structure failure due to fire was the collapse of two World Trade Center towers on September 11, 2001. The towers were hit by big passenger airplanes. A tremendous impact was inflicted on them, but they did not collapse immediately. The jet fuel started, however, intense fires and when the steel of some decisive members had reached critical temperatures a progressive collapse was initiated. For calculation methods on steel structures see Chap. 13.

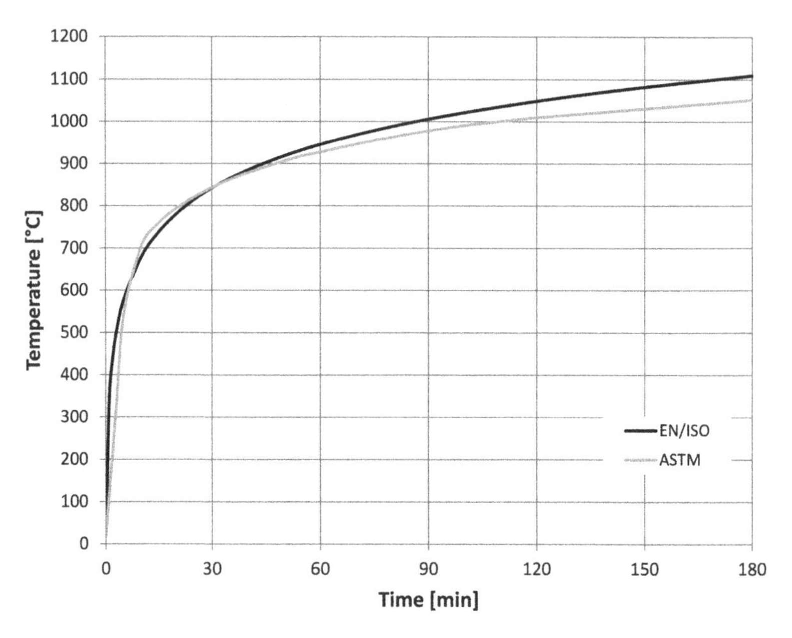

Fig. 1.9 The standard time–temperature curves according to EN 1363-1 or ISO 834 and ASTM E-119

Concrete also looses strength and stiffness at high temperature, see, e.g. Eurocode 2 [6] or [4]. Concrete has, however, a relatively low thermal conductivity and a high density and high specific heat capacity as well, i.e. a low thermal diffusivity. Although the temperature therefore rises slowly in concrete structures, it is important to assure that the steel reinforcement bars are not too near fire-exposed surfaces to avoid that their temperature reaches critical levels. See calculation methods in Chap. 14. An often more severe problem is the tendency of concrete to spall explosively when exposed to high temperature. In particularly high strength concrete qualities are prone to spall which is of great concern, for example, when designing linings of road and railway tunnels where fire temperatures may be extremely high and where a collapse may have devastating consequences in terms of life safety and protection of economic values.

Wood loses both strength and stiffness at elevated temperature. In addition it burns and chars gradually at a rate of about 0.5 mm/min when exposed to fire. The char layer then developed, however, protects the wood behind from being directly heated by the fire and thereby from quickly losing its load-bearing capacity. Timber structures therefore resist fire rather well and are in most cases left unprotected, see, e.g. Eurocode 5 [7]. In many cases structural timber members such as wall studs are protected from direct exposure by fire boards and can then resist fire for very long periods of time. For calculation methods see Chap. 15.

1.3.2 Polymers and Composite Materials

There are two main types of plastics materials, thermoplastics and thermosettings. They decompose differently when exposed to heat. Thermoplastics can soften with reverse changes of the material, while thermosetting materials are infusible and cannot undergo any simple phase changes. They do not have a fluid state.

Many thermoplastics and thermosetting materials form chars when decomposed by heat. This char is in general a good insulator and can protect the underlying virgin material from heat and slow down the decomposition process.

Polymers or plastics possess different hazards in fires depending on their physical constitution and chemical composition. In general, foamed plastics with low density and thin plastic objects ignite more easily and burn more vigorously than more dense and thick plastics. The fire properties of an object do not only depend on its chemical composition but also on the shapes and configurations. Thus a thin layer of a material ignites more easily when underlaid by a low density insulating material than by a more dense material. Below some characteristic are given of some commercially important polymers.

The thermal stability of *polyolefins* such as polyethylene and polypropylene depends on branching of the molecule chains, with linear polymers most stable and polymers with branching less stable. *Polyvinyl chloride* (*PVC*) has in general good fire properties as the chloride works as a flame retardant agent. However, the hydrochloride HCl, which is generated while burning, is irritating and toxic and can impede the evacuation from a fire. In addition, it forms hydrochloride acid when in contact with water and can therefore cause severe corrosion problems even long after a fire incident. *Polyurethanes* (*PU*) contain nitrogen and forms very toxic products such as hydrogen cyanide and isocyanides when burning. PVC and PU do also generate very dense smoke which can hamper escape possibilities.

Composite materials consisting of a polymer and reinforcing fibres (typically, glass, carbon or aramide fibres), also called "Fibre reinforced plastics (FRP)", have become increasingly used in many areas of construction, such as airplanes, helicopters and high-speed crafts, due to the high strength/weight ratio. These materials are also chemically very resistant and do not corrode or rust. They are, however, combustible and as they are often meant to replace non-combustible materials such as steel or other metals they could introduce new fire hazards.

1.3.3 Measurements of Material Properties

Material properties may be obtained from small scale laboratory test or derived from large scale fire test experiences. Small scale tests are in general the most accurate and cheap tests, but they are usually made for inert materials in room temperature. Therefore such data are relatively easy to find. In FSE, however, material data are needed at elevated temperature when these change, and in addition materials may undergo physical (e.g. vaporization of water) as well as chemical transformations (e.g. phase changes and pyrolysis). Then the small scale methods are generally unsuitable to consider these kinds of non-linear effects. In practice therefore thermal properties are often determined by "curve-fitting", i.e. measured temperatures are compared with calculated, and then input parameters are altered until measured and calculated data match as well as possible. In this way large scale non-linear effects may be considered. However, this kind of approach has the disadvantage that the results are valid only for the type of exposure being used to determine the data.

There are a number of techniques to measure thermal properties in small scale, each of them suitable for a limited range of materials, depending on thermal properties and temperature level, see, e.g. [8]. However, only a few of the measuring techniques can be used at high temperature levels relevant for fire conditions. They can be divided into steady-state and transient techniques.

The steady-state techniques perform the measurements when the material is in complete equilibrium. Disadvantages of these techniques are that it generally takes a long time to reach the required equilibrium and that at low temperature the measurements are influenced by moisture migration. For moist materials such as concrete, it is therefore often preferable to determine the apparent conductivity or thermal diffusivity with transient techniques. These techniques perform the measurements during a process of small temperature changes and can be made relatively quickly.

The guarded hot-plate is the most common steady-state method for building materials with a relatively low thermal conductivity. It is quite reliable at moderate temperatures up to about 400 °C.

As transient thermal processes dominate in FSE, the thermal diffusivity, a measure of the speed at which temperature is propagating into a material, is the most interesting parameter. It is naturally best measured with transient methods. One of the most interesting techniques is the transient plane source method (TPS). In this method a membrane, the TPS sensor, is located between two specimens halves and acts as heater as well as a temperature detector, see Fig. 1.10. By using this technique, thermal diffusivity, heat conductivity and volumetric specific heat can be obtained simultaneously for a variety of materials such as metals, concrete, mineral wool and even liquids and films [9].

Fig. 1.10 The TPS sensor placed between two pieces of a concrete specimen to measure thermal properties

Chapter 2
Steady-State Conduction

In one dimension in the x-direction the rate of heat transfer or heat flux is expressed according to *Fourier's law* as outlined in Sect. 1.1.

$$\dot{q}''_x = -k \cdot \frac{dT}{dx} \tag{2.1}$$

where k is the thermal conductivity. For simplicity the mathematical presentation of the heat transfer phenomena is here in general made for one-dimensional cases only. Corresponding presentations in two and three dimensions can be found in several textbooks such as [1, 2].

Under steady-state conditions the heat flux is independent of x, i.e. the derivative of $\dot{q}''_x$ is zero and we get

$$\frac{d}{dx}\left(k \cdot \frac{dT}{dx}\right) = 0 \tag{2.2}$$

The corresponding equation for cylinders with temperature gradients in the radial direction only is

$$\frac{1}{r}\frac{d}{dr}\left(k \cdot r\frac{dT}{dr}\right) = 0 \tag{2.3}$$

where r is the radius. Solutions for steady-state cases are found in Sect. 2.2.

U. Wickström, *Temperature Calculation in Fire Safety Engineering*,
DOI 10.1007/978-3-319-30172-3_2

2.1 Plane Walls

Consider a plane wall having surface temperatures T_1 and T_2. Figure 2.1 shows the temperature distribution under steady-state conditions which means the heat flux is constant across the plate. Figure 2.1a shows the temperature distribution when the heat conductivity is constant, i.e. the second derivative of the temperature is zero according to Eq. 2.2 and thus the temperature distribution becomes linear. Figure 2.1b shows the temperature distributions in structure with two layers of materials with different conductivities. The material to the left has the lower conductivity. Figure 2.1c indicates the temperature distribution when the conductivity is increasing with temperature. The temperature gradient is higher where the temperatures are lower and thereby the conductivity. This is particular the case for insulating materials where the conductivity increases considerably at elevated temperatures.

The rate of heat conducted per unit area $\dot{q}''$ through a wall, see Fig. 2.2, is proportional to the thermal conductivity of the wall material times the temperature difference ΔT between the wall surfaces divided by the wall thickness L, and according to *Fourier's law* (c.f. Eq. 2.1)

$$\dot{q}'' = k \cdot \frac{\Delta T}{L} = k \cdot \frac{(T_1 - T_2)}{L} \tag{2.4}$$

In an electric circuit analogy, this case can be illustrated according to Fig. 2.2. The heat flow through the wall over an area A may then be written as

$$\dot{q}'' = (T_1 - T_2)/R_k \tag{2.5}$$

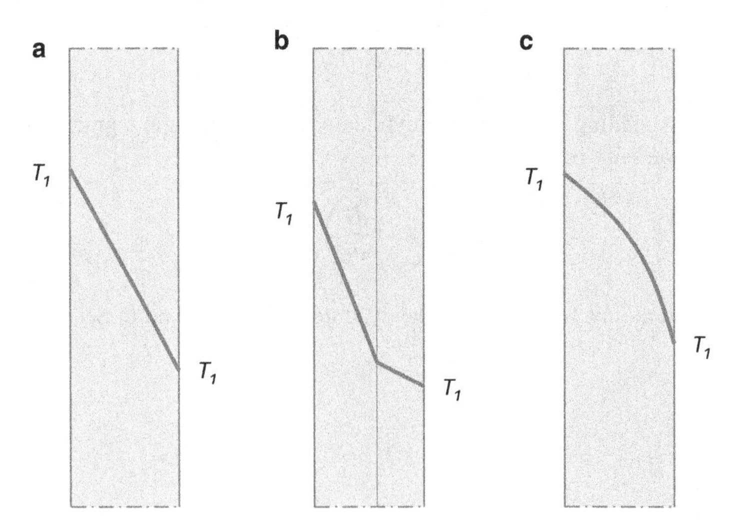

Fig. 2.1 Steady-state temperature distribution in a plane wall. (**a**) Constant conductivity, (**b**) two materials with a low and high conductivity and (**c**) conductivity increasing with temperature

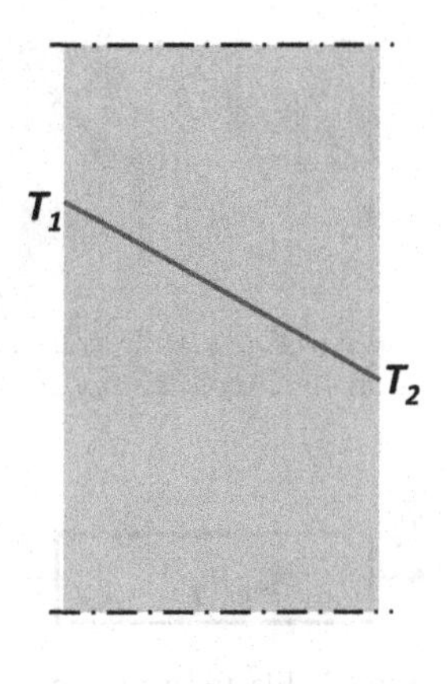

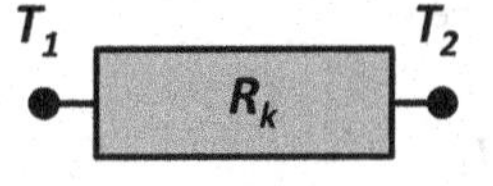

Fig. 2.2 One-dimensional steady-state thermal conduction. Linear temperature distribution across a wall and an electric analogy of one-dimensional heat flux. The thermal resistance $R = L/k$

where the *thermal resistance of the solid* then can be identified as

$$R_k = \frac{L}{k} \tag{2.6}$$

The electric analogy may also be used for more complex problems involving both series and parallel thermal resistance. A typical problem is a wall consisting of several layers, see Fig. 2.3.

The total thermal resistance R_{tot} between the inside and outside surfaces may then be written as:

$$R_{tot} = R_1 + R_2 + R_3 = \frac{L_1}{k_1} + \frac{L_2}{k_2} + \frac{L_3}{k_3} \tag{2.7}$$

and the heat flux $\dot{q}''$ through the assembly from the inside to the outside may be written as:

$$\dot{q}'' = \frac{\Delta T}{R_{tot}} = \frac{T_0 - T_\mathrm{i}}{R_{tot}} \tag{2.8}$$

The temperature T_{1-2} at the interface between material 1 and 2 may be written as

$$T_{1-2} = T_i - \frac{R_1 \left(T_i - T_o\right)}{R_{tot}} = \frac{(R_2 + R_3) \cdot T_i + R_1 \cdot T_o}{R_{tot}} \tag{2.9}$$

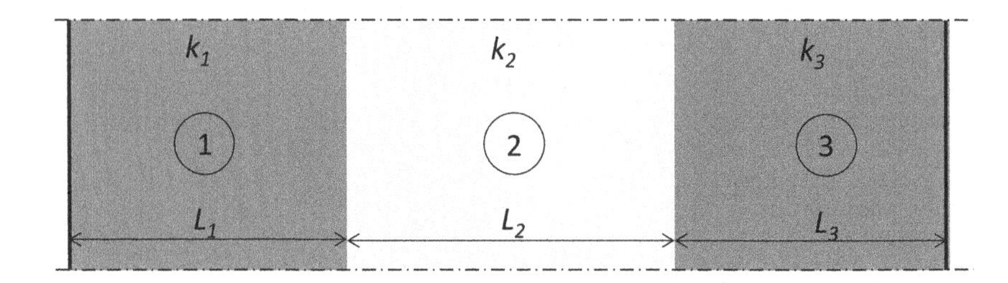

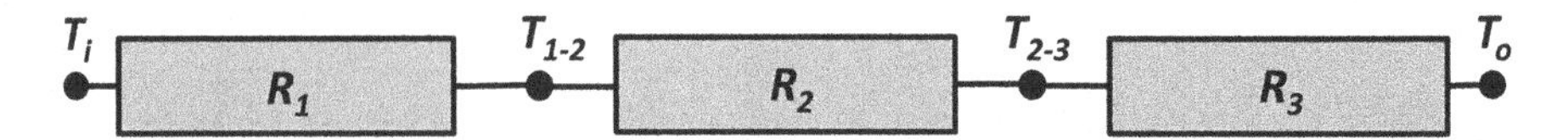

Fig. 2.3 Electric circuit analogy of one-dimensional heat transfer across a wall consisting of three layers

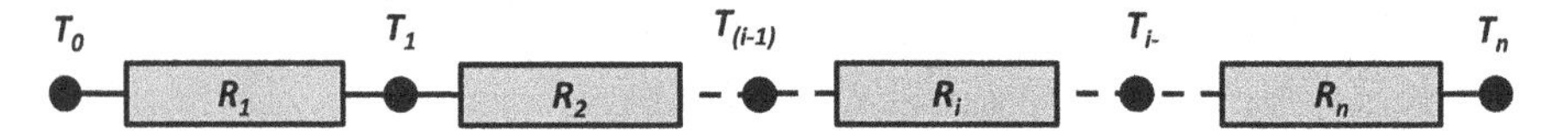

Fig. 2.4 Electric circuit analogy of one-dimensional heat transfer across a wall consisting of several layers

In a general way the total thermal resistance of an assembly thermally modelled as shown in Fig. 2.4 may be obtained as the sum of the components

$$R_{tot} = \sum_{j=1}^{n} R_j \tag{2.10}$$

and the heat flux $\dot{q}''$ through can be calculated as

$$\dot{q}'' = \frac{T_0 - T_n}{R_{tot}} \tag{2.11}$$

and the temperature at an interface i as shown in Fig. 2.4 may be calculated as

$$T_i = T_0 + \frac{\sum_{j=1}^{i} R_j}{R_{tot}} (T_n - T_0) = \frac{T_n \cdot \sum_{j=1}^{i} R_j + T_0 \cdot \sum_{j=i+1}^{n} R_j}{R_{tot}} \tag{2.12}$$

Example 2.1 A wall consists of 20 mm wood panel, 100 mm fibre insulation and 12 mm gypsum board with conductivities equal to 0.14, 0.04 and 0.5 W/(m^2 K), respectively. The wood outer surface has a constant temperature of 75 °C and the inner gypsum board surface a temperature of 15 °C. Calculate the temperatures at the insulation interface surfaces (T_1 and T_2).

Solution $R_{tot} = 0.020/0.14 + 0.1/0.04 + 0.012/0.5 = 0.143 + 2.5 + 0.024 = 2.67\text{W/K}$. Then $T_1 = 75 + \frac{0.143}{2.67} \cdot (15 - 75) = 71.8\,°\text{C}$ and $T_2 = 75 + \frac{0.143+2.5}{2.67} \cdot (15 - 75) = 15.6\,°\text{C}$.

The presentation so far includes heat transfer in solids only with boundary conditions of the first kind, i.e. prescribed surface temperatures. In most cases in fire protection engineering, however, the boundary condition between a surrounding fluid/environment and a solid surface is specified as a boundary condition of the third kind. In the simplest form the boundary condition is then described by the *Newton's law of cooling*. It may be seen as a heat transfer condition for convection and it states that the heat transfer to a surface is directly proportional to the difference between the surrounding gas temperature T_g and the surface temperature T_s:

$$\dot{q}'' = h \cdot (T_g - T_s) \tag{2.13}$$

where the constant of proportionality factor h is the heat transfer coefficient. The *surface thermal resistance* R_h between the gas phase and the solid phase can then be written as

$$R_h = \frac{1}{h} \tag{2.14}$$

Thus for the case illustrated in Fig. 2.5, the total resistance between the gas phase on the left side and the surface on the right side of the wall may be written as

$$R_{tot} = \left(\frac{1}{h} + \frac{L}{k}\right) = R_h + R_k \tag{2.15}$$

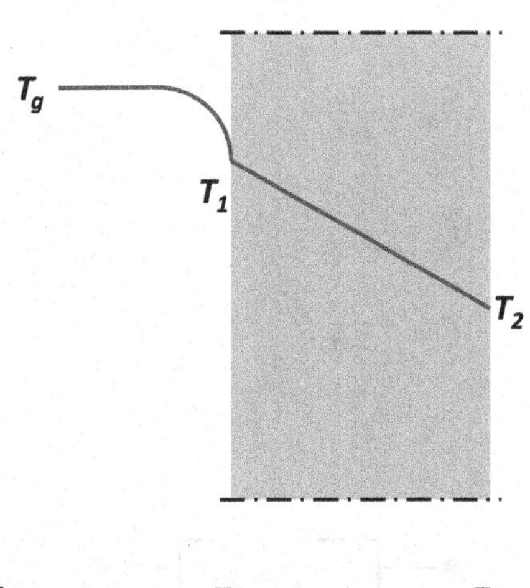

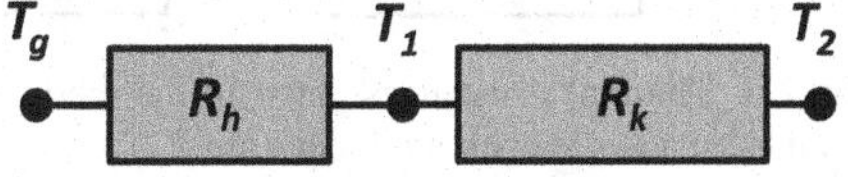

Fig. 2.5 An electric circuit analogy of one-dimensional heat transfer to a surface and through a wall with surface and solid thermal resistances

and the surface temperature T_1 may be written as a function of the gas temperature T_g and the temperature T_2 as

$$T_1 = \frac{R_h \cdot T_2 + R_k \cdot T_g}{R_h + R_k} \tag{2.16}$$

Example 2.2 Calculate the surface temperature T_I of a 12 mm wooden board if the gas temperature on the exposed side $T_g = 100$ °C and the temperature on the non-exposed side is $T_2 = 20$ °C. Assume the conductivity of the wood k = 0.2 W/(m K) and heat transfer coefficient $h = 5$ W/(m^2 K).

Solution Equation 2.15 yields $R = R_h + R_k = \left(\frac{1}{5} + \frac{0.012}{0.2}\right)(\mathrm{m}^2\,\mathrm{K})/\mathrm{W} = 0.2 + 0.06$ $(\mathrm{m}^2\,\mathrm{K})/\mathrm{W}$ and Eq. 2.16 yields $T_1 = \frac{0.2 \cdot 20 + 0.06 \cdot 100}{0.26}\,°\mathrm{C} = 38\,°\mathrm{C}$.

2.2 Cylinders

Cylinders often experience temperature gradients in the radial direction only, and may therefore be treated as one dimensional. The solid thermal resistance between the inner radius r_i and an arbitrary radius r in a cylinder (see Fig. 2.6) assuming constant heat conductivity may then be written as

$$R_k = \frac{\ln\left(r/r_i\right)}{2\pi k} \tag{2.17}$$

$$R_{hi} = \frac{1}{2\pi r_i h_i}$$

$$R_k = \frac{\ln(r_o/r_i)}{2\pi k}$$

$$R_{ho} = \frac{1}{2\pi r_o h_o}$$

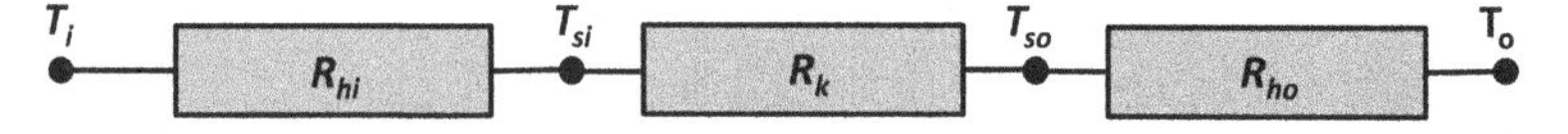

Fig. 2.6 Thermal resistances between the media with a temperature T_i inside a cylindrical pipe and the outside gas with a temperature T_o

where k is the thermal conductivity. The surface thermal resistance may be written as

$$R_h = \frac{1}{2\pi rh} \tag{2.18}$$

Hence the thermal resistance between the inner and outer gases or liquids of a pipe is obtained by summarizing the surface and solid resistances as indicated in Fig. 2.6, i.e. the total thermal resistance over a unit length is

$$R_{tot} = R_{hi} + R_k + R_{ho} = \frac{1}{2\pi} \cdot \left(\frac{1}{rh_i} + \frac{\ln\left({}^{r_o}/_{r_i} \right)}{k} + \frac{1}{r_o h_o} \right) \tag{2.19}$$

A uniform heat flux over a unit length of a pipe may then be calculated as

$$\dot{q}'_l = \frac{T_i - T_o}{R_{tot}} = \frac{2\pi \left(T_i - T_o \right)}{\frac{1}{r_i \cdot h_i} + \frac{\ln\left({}^{r_o}/_{r_i} \right)}{k} + \frac{1}{r_o \cdot h_o}} \tag{2.20}$$

The temperatures T_{is} at the inner surface can be obtained as

$$T_{is} = \frac{R_{hi} T_o + \left(R_k + R_{ho} \right) T_i}{R_{tot}} \tag{2.21}$$

and T_{os} at the outer surface as

$$T_{os} = \frac{\left(R_{hi} + R_k \right) T_o + R_{ho} T_i}{R_{tot}} \tag{2.22}$$

Example 2.3 Consider an insulated steel pipe with an outer coating as shown in Fig. 2.7 exposed to fire with a constant temperature of 800 °C. The temperature of

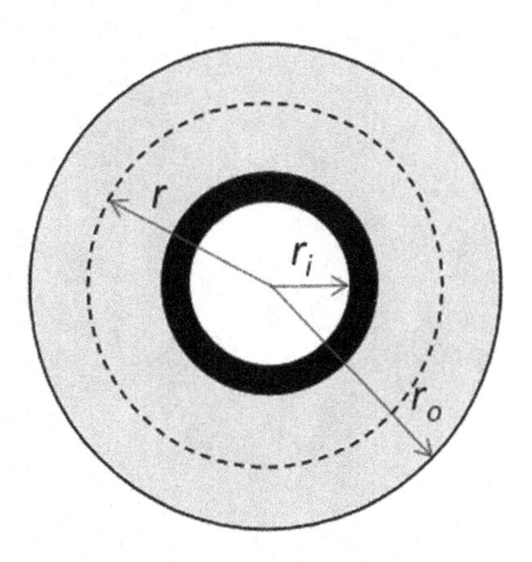

Fig. 2.7 Insulated steel pipe with an outer coating

the inside medium/fluid is 100 °C. The inner and outer radii of the steel pipe are 30 and 28 mm, respectively. The insulation is 50 mm thick and has a conductivity of 0.5 W/(m K). The inner heat transfer coefficient is 100 W/(m^2 K) and the outer 50 W/(m^2 K). Calculate the temperature of the steel pipe which is assumed to be constant along the radius.

Solution Calculating for a unit length. Equation 2.19 yields $R_{tot} = 0.057 + 0.312 + 0.040 = 0.409$ (m K)/W. Thus the inner (steel) temperature $T_i = \frac{0.057 \cdot 800 + (0.312 + 0.04) \cdot 100}{0.409} = 198\,°\mathrm{C}$.

Chapter 3
Unsteady-State Conduction

When a body is exposed to unsteady or transient thermal conditions, its temperature changes gradually, and if the exposure conditions remain constant it will eventually come to a new steady state or equilibrium. The rate of this process depends on the mass and thermal properties of the exposed body, and on the heat transfer conditions. As a general rule the lighter a body is (i.e. the less mass) and the larger its surface is, the quicker it adjusts to a new temperature level, and vice versa. The temperature development is governed by the heat conduction equation (Eq. 1.29) with the assigned boundary conditions. It can be solved analytically in some cases, see textbooks such as [1, 2], but usually numerical methods are needed. This is particular the case in fire protection engineering problems where temperature generally varies over a wide range, often several hundred degrees.

There are, however, some cases where analytical methods can be used. Two cases are of interest for both practical uses and basic understandings of the influence of material properties on their fire behaviour. On one hand, it is cases where bodies can be assumed to have *uniform temperature* such as in thin solids or in metals with a high conductivity. Then the approximation of *lumped-heat-capacity* can be applied. On the other hand, it is the case when a body can be assumed *semi-infinitely thick* for the time span considered. Then in particular the surface temperature can be estimated by analytical methods if the material properties are assumed constant. These two elementary cases will be considered in detail in the following two Sects. 3.1 and 3.2.

3.1 Lumped-Heat-Capacity

It is often assumed when calculating temperature in steel sections, protected as well as unprotected, that the temperature is uniform in the exposed body, see Sects. 13.3 and 13.4, respectively. It may also be applied when estimating temperature and time to ignition of thin materials such as curtain fabrics. A special case is the analysis of the temperature development of thermocouples and the definition of time constants

U. Wickström, *Temperature Calculation in Fire Safety Engineering*,
DOI 10.1007/978-3-319-30172-3_3

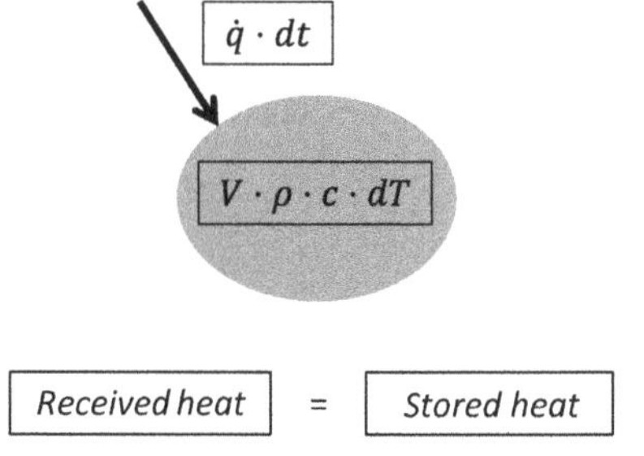

Fig. 3.1 The dynamic heat balance of a body over a time period is expressed as the heat received is equal to the heat stored according to Eq. 3.1

of these types of measuring devices, see Sect. 9.1. Numerical methods for calculating temperature when assuming lumped-heat-capacity are described in Sect. 7.1.

In this section a general presentation will be given assuming constant heat transfer coefficients, material properties and exposure levels. As only one temperature independent of position is calculated, this type of problems are *zero-dimensional.*

The received heat over a time interval dt is equal to the heat stored. The latter is proportional to the temperature rise of the body dT, see Fig. 3.1. Thus

$$\dot{q} \cdot dt = V \cdot \rho \cdot c \cdot dT \tag{3.1}$$

Hence the temperature rise rate dT/dt (the time derivative of the body temperature) vs. incident heat flow $\dot{q}\cdot$ or the incident heat flux $\dot{q}''$ can be obtained as

$$\frac{dT}{dt} = \frac{1}{V \cdot \rho \cdot c} \dot{q} = \frac{A}{V \cdot \rho \cdot c} \dot{q}'' \tag{3.2}$$

where A is exposed area, V volume, ρ density and c specific heat capacity. For thin plates exposed from one side V/A may be replaced by its thickness d

$$d = \frac{V}{A} \tag{3.3}$$

The heat flux $\dot{q}''$ to the body can be obtained in various ways depending on the boundary condition. It may be of the second or third kind, see Sect. 1.1.3. The first kind is trivial as a uniform temperature is assumed.

3.1.1 Prescribed Heat Flux: BC of the Second Kind

Given a prescribed heat flux $\dot{q}''$ (second kind of BC, see Sect. 1.1.3), the temperature rise $T - T_i$ as function of time may be obtained by integrating over time as

$$T - T_i = \frac{A}{V \cdot (c \cdot \rho)} \int_0^t \dot{q}'' \cdot dt \tag{3.4}$$

where T_i is the initial temperature. If $\dot{q}''$ remains constant over time

$$T = T_i + \frac{A \cdot \dot{q}'' \cdot t}{V \cdot (c \cdot \rho)} \tag{3.5}$$

Prescribed heat flux can rarely be assumed in fire protection engineering as the heat flux from the gas phase to a solid surface depends on the surface temperature which changes over time. Instead it is the third kind of BC that generally applies.

3.1.2 Prescribed Gas Temperature: BC of the Third Kind—And the Concept of Time Constant

More realistic and commonly assumed in FSE problems is that the heat transfer to a surface is proportional to the difference between the surrounding gas or fire temperature T_f and the body temperature as indicated in Fig. 3.2 (third kind of BC, see Sect. 1.1.3). The body having uniform temperature (lumped-heat-capacity) is here assumed to be of steel and its temperature is denoted T_{st}.

In case of *uninsulated* or *unprotected* bodies such as bare steel sections the surface thermal resistance is the only thermal resistance between the fire gases and the steel. Thus the heat flux can be written as

$$\dot{q}'' = h \cdot (T_f - T_{st}) = \frac{(T_f - T_{st})}{R_h} \tag{3.6}$$

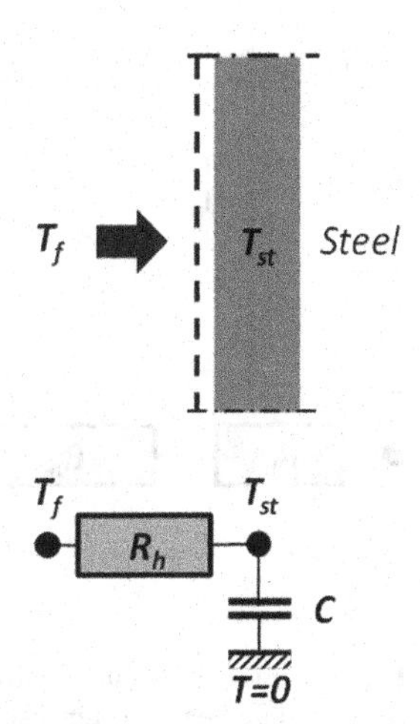

Fig. 3.2 Electric circuit analogy of an uninsulated steel section assumed to have uniform temperature (lumped-heat-capacity)

where h is the heat transfer coefficient and R_h the corresponding surface resistance which can be identified as

$$R_h = \frac{1}{h} \tag{3.7}$$

For an *insulated* or *protected* steel section, the heat resistance is the sum of the surface thermal resistance R_h and the solid resistance of the insulation R_{in}. (Notice that *insulated* and *protected* are used synonymously in this book). The total thermal resistance between the fire gases and the steel section is then

$$R_{h+in} = R_h + R_{in} \tag{3.8}$$

where

$$R_{in} = \frac{d_{in}}{k_{in}} \tag{3.9}$$

and where d_{in} and k_{in} are the insulation thickness and conductivity, respectively. Electric circuit analogies are shown in Fig. 3.3. In Fig. 3.3a the surface resistance is included while in Fig. 3.3b it is not. That assumption is made in many cases as the surface resistance is much smaller than the solid resistance, i.e. $R_h \ll R_{in}$, and may therefore be ignored (as suggested in, e.g. Eurocode 3 [3]).

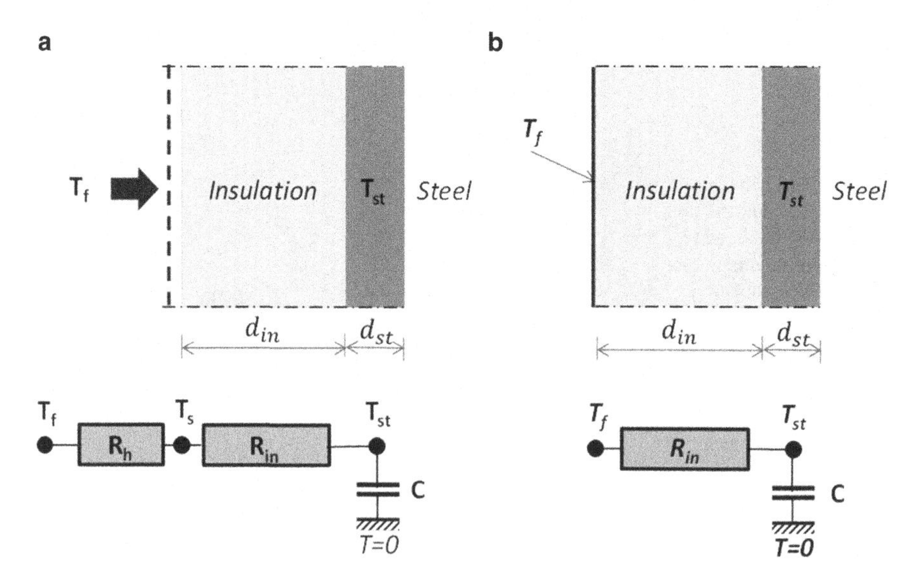

Fig. 3.3 Electric circuit analogy of an insulated steel section treated as a lumped-heat-capacity. (**a**) Including heat transfer resistance. (**b**) Neglecting heat transfer resistance

For a unit area ($A = 1$) Eq. 3.2 may be written as

$$\dot{q}'' = d_{st} \cdot \rho_{st} \cdot c_{st} \cdot \frac{dT_{st}}{dt} = C_{st} \cdot \frac{dT_{st}}{dt} \tag{3.10}$$

where d_{st} is the thickness and C is the heat capacitance per unit exposed area (see Table 1.1), i.e.

$$C_{st} = d_{st} \cdot \rho_{st} \cdot c_{st} \tag{3.11}$$

Now by combining Eq. 3.2 with Eq. 3.6

$$\frac{dT}{dt} = \frac{(T_f - T_{st})}{C_{st} \cdot R_{h+in}} \tag{3.12}$$

where alternatively Eq. 3.12 can be written as

$$\frac{dT}{dt} = \frac{1}{\tau}(T_f - T_{st}) \tag{3.13}$$

and where τ may be identified as the *time constant*:

$$\tau = C_{st} \cdot R_{h+in} \tag{3.14}$$

Then if the surrounding temperature T_f is constant and the time constant τ including the material and heat transfer parameters remains constant, Eq. 3.12 has an analytical solution

$$\frac{T_{st} - T_i}{T_f - T_i} = 1 - e^{-\frac{t}{\tau}} \tag{3.15}$$

where T_i is the initial temperature at time $t = 0$. The relation is shown in Fig. 3.4.

Notice that Eq. 3.15 may only be applied when constant material properties and surface resistances are assumed. That is, however, not so common in FSE and therefore must in most cases numerical solutions be used. More on numerical solutions will be shown in Sect. 7.1 and more on steel sections in Chap. 13.

3.1.2.1 Gas Temperature Varying with Time

Equation 3.15 may be applied only to a sudden change of the exposure temperature to a new constant value. When the exposure temperature varies with time, superposition techniques may be applied as outlined in Sect. 7.2. As an example the temperature of a steel section when assuming a constant time constant τ can be obtained by superposition as

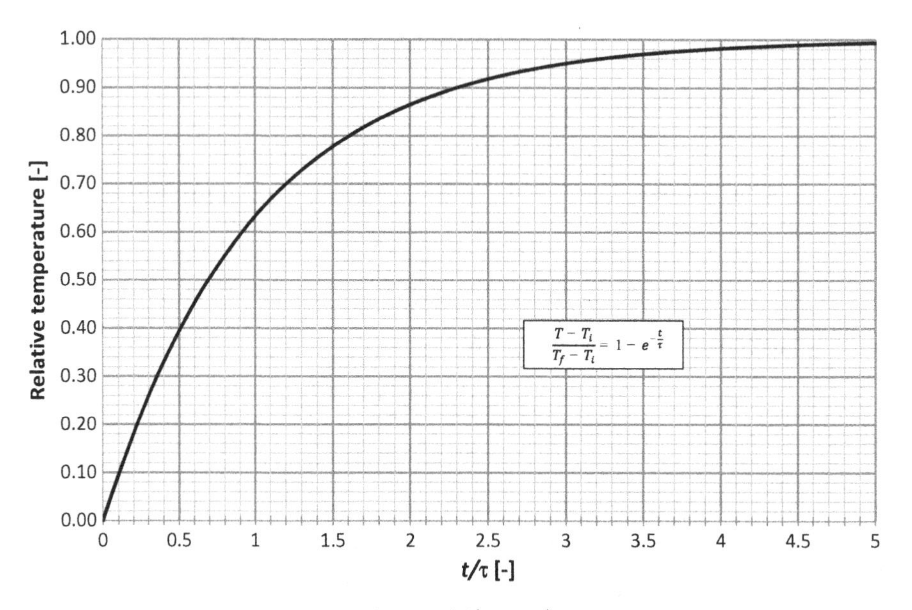

Fig. 3.4 The relative temperature rise $(T_{st} - T_i)/(T_f - T_i)$ of a body with uniform temperature vs. dimensionless time t/τ according to Eq. 3.15

$$T_{st} = \frac{1}{\tau} e^{-t/\tau} \int_0^t T_f\,(\xi) e^{\xi/\tau} d\xi + e^{-t/\tau} \cdot T_i \tag{3.16}$$

The integral of Eq. 3.16 can be solved analytically in some cases depending on the analytical expression of the gas temperature T_f as a function of time. For instance, when the fire temperature rises linearly with time as

$$\Delta T_f = a \cdot t \tag{3.17}$$

the steel temperature rise becomes

$$\Delta T_{st} = a \cdot t \left[1 - \frac{\tau}{t} \left(1 - e^{-\frac{t}{\tau}} \right) \right] \tag{3.18}$$

or in dimensionless format

$$\frac{\Delta T_{st}}{a \cdot \tau} = \frac{t}{\tau} \left[1 - \frac{\tau}{t} \left(1 - e^{-\frac{t}{\tau}} \right) \right] \tag{3.19}$$

Figure 3.5 shows that the steel temperature rise asymptotically approaches a temperature $(\Delta T_f - a \cdot \tau)$.

Another example where the integral of Eq. 3.16 can be solved analytically is shown in Sect. 13.3.2 dealing with fire insulated steel sections exposed to parametric fires according to Eurocode EN 1991-1-2. The fire temperature curve is then

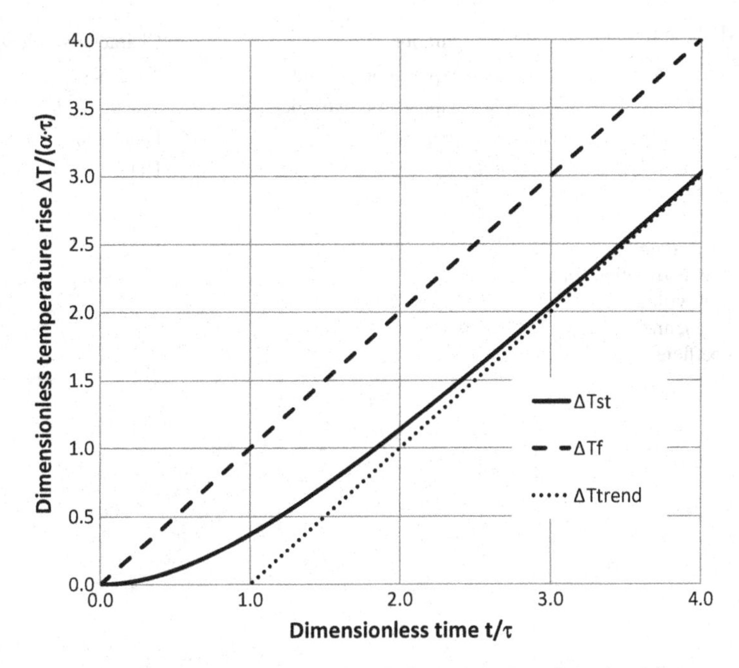

Fig. 3.5 Dimensionless temperature rise of a body exposed to linearly rising temperature vs. dimensionless time according to Eq. 3.19

expressed as a sum of exponential terms which allows the steel temperature to be calculated analytically according to Eq. 3.16. See also the Sect. 9.1 on the response of thermocouples.

3.1.3 Conditions for Assuming Lumped-Heat-Capacity

The assumption of lumped-heat-capacity or uniform body temperature is an approximation which may be applied when the internal thermal resistance by conduction is low in comparison to the heat transfer resistance in the case of a non-insulated body, i.e. the Biot number defined as

$$Bi = \frac{L/k}{1/h} = \frac{hL}{k} \tag{3.20}$$

is less than 0.1, see, e.g. [2]. Here L is a characteristic length of the body studied as shown in Table 3.1, and k is the conductivity. Figure 3.6 shows the effect of various Biot numbers of plane body under steady-state conditions.

Table 3.1 Examples of characteristic lengths L

Configuration	Characteristic length, L
Plate exposed on one side	Thickness
Plate exposed on two sides	Thickness/2
Long cylinder	Diameter/4
Sphere	Diameter/6

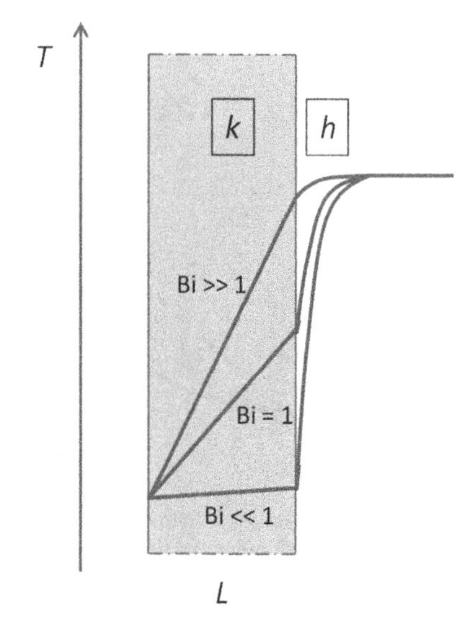

Fig. 3.6 The state temperature distribution for a plane wall with a conductivity k and a heat transfer coefficient h

In fire protection engineering lumped-heat-capacity is often assumed for steel sections. This is in particular appropriate when considering temperature across the thickness of a web or flange. The temperatures on the two sides of a metal sheet are then by and large equal. Temperatures along the plane of a web or flange may, however, vary considerably depending on the boundary conditions, see Chap. 13. The criterion given by Eq. 3.20 is based on steady-state conditions but applies for steel of thicknesses in the order of 10 mm except for the first few minutes which are usually not of interest in FSE.

Example 3.1 A 2-mm-thick steel plate with an initial temperature $T_i = 20$ °C is suddenly exposed on both sides to a gas temperature of 500 °C. Assume a constant heat transfer coefficient $h = 20$ W/(m^2 K) and the steel properties $c = 460$ Ws/(kg K) and $\rho = 7850$ kg/m^3.

(a) Calculate the thermal time constant of the steel plate.
(b) Calculate the temperature of the steel plate after 5 min.

Solution (a)
$$\tau = \frac{dc\rho}{h} = \frac{0.5 \cdot 0.002 \cdot 460 \cdot 7850}{20} = 181 \text{ s}$$

(b) At time t = 300 s the dimensionless time $t/\tau = 300/181 = 1.66$ and from Eq. 3.15 or Fig. 3.4 the steel temperature is calculated as $T_{st} = 20 + (500 - 20) \cdot \left(1 - e^{-1.66}\right) = 20 + 480 \cdot 0.81 = 409\ °\text{C}$.

Example 3.2 A 5-mm-thick steel bulkhead is suddenly exposed to a fire with a constant temperature of $T_f = 1000\ °\text{C}$. It is insulated on the fire-exposed side and uninsulated on the non-fire-exposed side. The insulation thickness $d = 100$ mm and its conductivity $k = 0.07$ W/(m K). The heat transfer coefficient on the non-fire-exposed side $h = 5$ W/(m^2 K). Assume the surface heat resistance on the fire-exposed side is negligible and the steel properties $c = 460$ Ws/(kg K) and $\rho = 7850$ kg/m^3, and the ambient temperature and the initial temperature $T_\infty = T_i = 20\ °\text{C}$.

(a) What is the ultimate steel temperature T_{st}^{ult}?
(b) What is the time constant τ of the bulkhead?
(c) What is the steel temperature after 60 min?

Solution (a) The heat balance equation of the steel bulkhead can be written as $\frac{k}{d}\left(T_f - T_{st}\right) + h(T_\infty - T_{st}) = d \cdot c \cdot \rho \frac{dT_{st}}{dt}$. The ultimate steel temperature is obtained when $\frac{dT_{st}}{dt} = 0$ and then $T_{st}^{ult} = \frac{\frac{k}{d}T_f + hT_\infty}{\frac{k}{d} + h} = \frac{\frac{0.07}{0.1} \cdot 1000 + 5 \cdot 20}{\frac{0.07}{0.1} + 5} = \frac{800}{5.7} = 140\ °\text{C}$.

(b) The heat balance equation can be reorganized and written as $\left\{\frac{k}{d} \cdot T_f + h \cdot T_\infty\right\} - \left\{\frac{k}{d} + h\right\} T_{st} = d \cdot c \cdot \rho \frac{d \cdot T_{st}}{dt}$ and $\frac{dT_{st}}{dt} = \frac{\left\{\frac{k}{d} \cdot T_f + h \cdot T_\infty\right\}}{d \cdot c \cdot \rho} - \frac{\left\{\frac{k}{d} + h\right\}}{d \cdot c \cdot \rho} T_{st} = \frac{T_{st}^{ult} - T_{st}}{\frac{d \cdot c \cdot \rho}{\frac{k}{d} + h}}$. Now the time constant can be identified (compare with Eq. 3.13) as $\tau = \frac{d \cdot c \cdot \rho}{\frac{k}{d} + h} = \frac{0.005 \cdot 460 \cdot 7850}{5.7} = 3170$ s.

(c) At time $t = 3600$ s the dimensionless time $t/\tau = 3600/3170 = 1.14$ and from Eq. 3.15 or Fig. 3.4 $T_{st} = T_i + \left(T_{st}^{ult} - T_i\right) \cdot \left(1 - e^{-\frac{t}{\tau}}\right) = 20 + 120 \cdot 0.68 = 101\ °\text{C}$.

3.2 Semi-infinite Solids

Cases and scenarios in FSE are often short in time. Therefore only the surface and the top layer of a solid will be involved in fire phenomena such as ignition and flame spread. In such cases a solid may be assumed semi-infinite as its surface will not thermally be influenced by the limited depth of the exposed surface layer. Even concrete elements exposed to fires of an hour's duration any temperature rise beyond 200 mm from the surface is insignificant, and when estimating temperature of reinforcement bars near the exposed surface may even slender structures be considered as semi-infinite.

Whether a body can be treated as semi-infinite depends on time of consideration and thickness of the exposed body or more precise the exposed layer in case of composites. The longer a body is analysed, the thicker it must be to be treated as semi-infinite. As a general rule the change of temperature at one point influences the temperature only within a distance δ proportional to the square root of the thermal diffusivity $\alpha = k/(c \cdot \rho)$ according to Eq. 1.28 multiplied by the time t:

$$\delta < 3\sqrt{\alpha t} \tag{3.21}$$

The coefficient "3" here is an arbitrary value depending on accuracy, see Sect. 3.2.1.1 on thermal penetration depth.

Consider a semi-infinite body initially at a constant temperature T_i. Three kinds of boundary conditions can be identified (c.f. Sect. 1.1.3 and Fig. 3.7):

(a) The surface temperature changes suddenly to a new constant value. The internal temperature distribution (Eq. 3.22) and the surface flux (Eq. 3.23) can then be calculated (first kind of BC).
(b) The surface receives suddenly a constant heat flux. The surface temperature can then be calculated according to Eq. 3.29 (second kind of BC).
(c) The surface is suddenly exposed to a constant gas temperature and the heat flux to the surface is proportional to the temperature difference between the gas temperature and the surface temperature. The surface temperature can then be calculated according to Eq. 3.35. For internal temperatures closed form solutions can be found in textbooks such as [1, 2] (third kind of BC).

For more details on the three types of boundary conditions see Sect. 1.1.3.

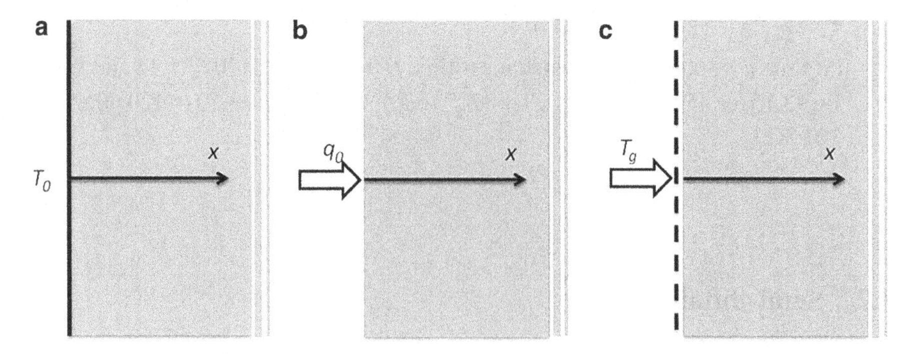

Fig. 3.7 Semi-infinite bodies with the three kinds of BC. See Sects. 3.2.1–3.2.3, respectively. (**a**) First kind. (**b**) Second kind. (**c**) Third kind

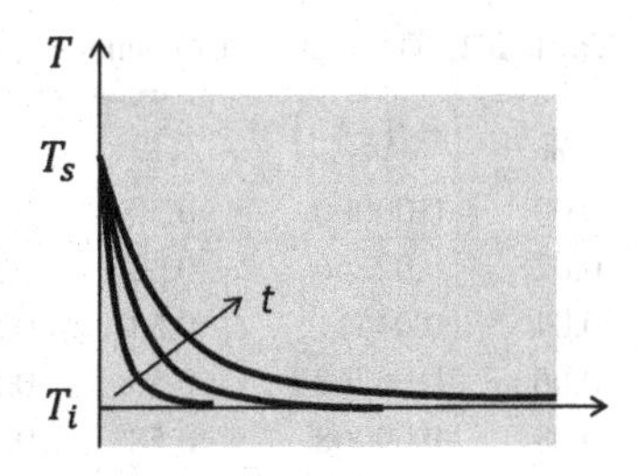

Fig. 3.8 Semi-infinite solid with an initial temperature T_i where the surface temperature changes suddenly to T_s

3.2.1 Constant Surface Temperature: First Kind of BC

The surface temperature of a semi-infinite body is suddenly changed from its initial temperature T_i to T_s. Then temperature profiles as indicated in Fig. 3.8 develop. According to Sect. 1.1.3 this is a first kind of boundary condition. The longer times t the further into the body the temperature rise goes. Mathematically the temperature distribution may be written as

$$\frac{T(x,t) - T_i}{T_s - T_i} = 1 - erf\left(\frac{x}{2\sqrt{\alpha t}}\right) \equiv erfc\left(\frac{x}{2\sqrt{\alpha t}}\right) \tag{3.22}$$

where x is the distance to the surface and α the *thermal diffusivity*, see Eq. 1.28. The function *erf* is called the Gauss error function and *erfc* is its complimentary function. The error function is tabulated in Table 3.2 and both *erf* and *erfc* are shown in Fig. 3.9 as functions of the dimensionless parameter group $x/(2\sqrt{\alpha t})$.

The heat flux at the boundary $\dot{q}''_0$ becomes

$$\dot{q}''_0 = -k\left(\frac{dT}{dx}\right) = \frac{1}{\sqrt{\pi}}(T_s - T_i)\sqrt{\frac{k \cdot \rho \cdot c}{t}} \tag{3.23}$$

This equation shows that the heat flux at a given time and temperature rise is proportional to the square root of the parameter group $(k \cdot \rho \cdot c)$. This is often referred to as the *thermal inertia* of the material. (In some literature the thermal inertia is defined as $\sqrt{k \cdot \rho \cdot c}$). The thermal inertia of a material has a great influence on its ignition and flame spread properties which will be discussed further in Sect. 3.2.3 on third kind of BC.

3.2.1.1 Temperature Penetration Depth

The rate at which a temperature change diffuses into a body when exposed to heating conditions depends for a semi-infinite body on the thermal diffusivity α as defined in Eq. 1.28. Therefore it takes a relatively long time for a temperature rise at the surface to penetrate into a material with a low conductivity k and/or a high

Table 3.2 The Gauss error function, *erf*

$\frac{x}{2\sqrt{\alpha t}}$	$erf\left(\frac{x}{2\sqrt{\alpha t}}\right)$	$\frac{x}{2\sqrt{\alpha t}}$	$erf\left(\frac{x}{2\sqrt{\alpha t}}\right)$	$\frac{x}{2\sqrt{\alpha t}}$	$erf\left(\frac{x}{2\sqrt{\alpha t}}\right)$	$\frac{x}{2\sqrt{\alpha t}}$	$erf\left(\frac{x}{2\sqrt{\alpha t}}\right)$
0.00	0.00000	0.50	0.52050	1.00	0.84270	1.50	0.96611
0.02	0.02256	0.52	0.53790	1.02	0.85084	1.52	0.96841
0.04	0.04511	0.54	0.55494	1.04	0.85865	1.54	0.97059
0.06	0.06762	0.56	0.57162	1.06	0.86614	1.56	0.97263
0.08	0.09008	0.58	0.58792	1.08	0.87333	1.58	0.97455
0.10	0.11246	0.60	0.60386	1.10	0.88021	1.60	0.97635
0.12	0.13476	0.62	0.61941	1.12	0.88679	1.62	0.97804
0.14	0.15695	0.64	0.63459	1.14	0.89308	1.64	0.97962
0.16	0.17901	0.66	0.64938	1.16	0.89910	1.66	0.98110
0.18	0.20094	0.68	0.66378	1.18	0.90484	1.68	0.98249
0.20	0.22270	0.70	0.67780	1.20	0.91031	1.70	0.98379
0.22	0.24430	0.72	0.69143	1.22	0.91553	1.72	0.98500
0.24	0.26570	0.74	0.70468	1.24	0.92051	1.74	0.98613
0.26	0.28690	0.76	0.71754	1.26	0.92524	1.76	0.98719
0.28	0.30788	0.78	0.73001	1.28	0.92973	1.78	0.98817
0.30	0.32863	0.80	0.74210	1.30	0.93401	1.80	0.98909
0.32	0.34913	0.82	0.75381	1.32	0.93807	1.82	0.98994
0.34	0.36936	0.84	0.76514	1.34	0.94191	1.84	0.99074
0.36	0.38933	0.86	0.77610	1.36	0.94556	1.86	0.99147
0.38	0.40901	0.88	0.78669	1.38	0.94902	1.88	0.99216
0.40	0.42839	0.90	0.79691	1.40	0.95229	1.90	0.99279
0.42	0.44747	0.92	0.80677	1.42	0.95538	1.92	0.99338
0.44	0.46623	0.94	0.81627	1.44	0.95830	1.94	0.99392
0.46	0.48466	0.96	0.82542	1.46	0.96105	1.96	0.99443
0.48	0.50275	0.98	0.83423	1.48	0.96365	1.98	0.99489

volumetric heat capacity $(c \cdot \rho)$. Thus any temperature change diffuses, e.g. much faster in a concrete than in a steel.

For the idealized case of a semi-infinite body at a uniform initial temperature T_i where the surface temperature momentarily is changed to a constant level of T_s, the temperature rise $(T - T_i)$ inside the body at a depth x at a time t may be written as a function of the normalized group

$$\eta = x/\left[2\sqrt{(\alpha \cdot t)}\right] \tag{3.24}$$

where the thermal diffusivity $\alpha = k/(c \cdot \rho)$ according to Eq. 1.28 is assumed constant. The relative temperature rise may then be written as:

$$\frac{(T - T_i)}{(T_s - T_i)} = erfc\ (\eta) = 1 - erf(\eta) \tag{3.25}$$

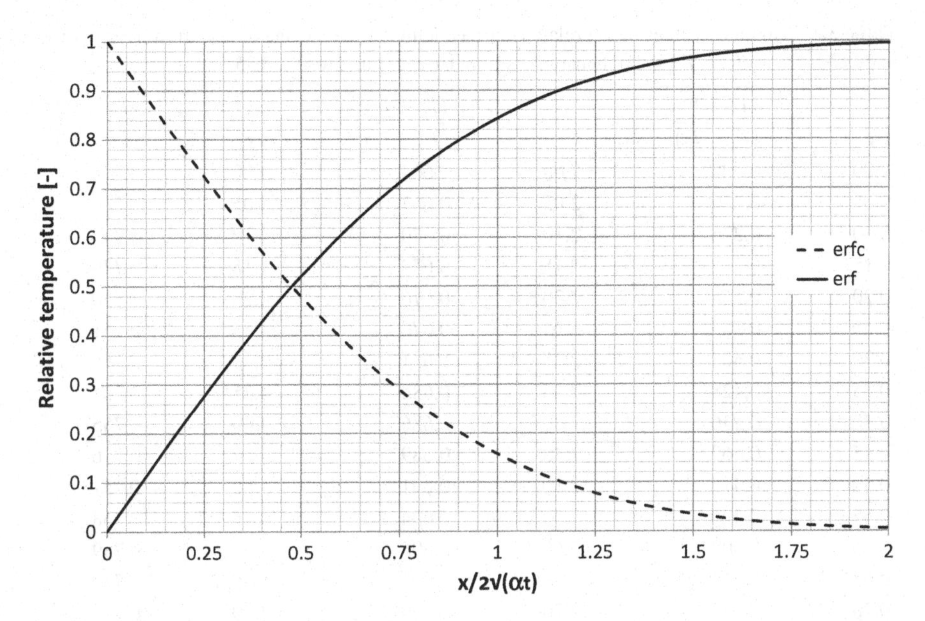

Fig. 3.9 The Gauss error-function $erf\left(x/\left(2\sqrt{\alpha t}\right)\right.$ and the Gauss complimentary error-function $erfc\left(x/\left(2\sqrt{\alpha t}\right)\right)$

For the Gauss error function see Fig. 3.9 and Table 3.3. Note that for values of $\eta \geq 1.4$ the relative temperature rise is less than 5 %, and for $\eta \geq 1.8$ it is less than 1 %. This can be interpreted as the temperature penetration depth δ which can be derived from Eq. 3.24 by solving for x. Thus the 5 % penetration depth is

$$\delta_{0.05} = 2.8\sqrt{\alpha \cdot t} \tag{3.26}$$

and the corresponding 1 % is

$$\delta_{0.01} = 3.6\sqrt{\alpha \cdot t} \tag{3.27}$$

Hence a sudden temperature rise at the surface will penetrate in 30 min about 0.14 m into a concrete structure and about four times longer (0.54 m) into or along a steel structure. Constant material properties are then assumed according to Table 1.2.

Example 3.3 The surface temperature of a thick concrete wall with an initial temperature of 0 °C rises suddenly to 1000 °C.

(a) What is the 1 % thermal penetration depth $\delta_{0.01}$ after 15 min?
(b) What is the temperature T at that point after 60 min?

Assume constant concrete properties according to Table 1.2, i.e. $c = 900$ Ws/(kg K), $\rho = 2300$ kg/m^3 and $k = 1.5$ W/(m K).

Table 3.3 Tabulated values of the relative surface temperature change of a semi-infinitely thick body $(T_s - T_i)/(T_g - T_i)$ vs. dimensionless time t/τ according to Eq. 3.35

t/τ	$\frac{(T_s - T_i)}{(T_g - T_i)}$	t/τ	$\frac{(T_s - T_i)}{(T_g - T_i)}$	t/τ	$\frac{(T_s - T_i)}{(T_g - T_i)}$
0.00	0.000	2.00	0.664	4.00	0.745
0.05	0.210	2.05	0.667	4.05	0.746
0.10	0.276	2.10	0.670	4.10	0.747
0.15	0.322	2.15	0.673	4.15	0.749
0.20	0.356	2.20	0.676	4.20	0.750
0.25	0.384	2.25	0.678	4.25	0.751
0.30	0.408	2.30	0.681	4.30	0.752
0.35	0.428	2.35	0.684	4.35	0.753
0.40	0.446	2.40	0.686	4.40	0.755
0.45	0.462	2.45	0.689	4.45	0.756
0.50	0.477	2.50	0.691	4.50	0.757
0.55	0.490	2.55	0.694	4.55	0.758
0.60	0.502	2.60	0.696	4.60	0.759
0.65	0.513	2.65	0.698	4.65	0.760
0.70	0.523	2.70	0.700	4.70	0.761
0.75	0.533	2.75	0.703	4.75	0.763
0.80	0.542	2.80	0.705	4.80	0.764
0.85	0.550	2.85	0.707	4.85	0.765
0.90	0.558	2.90	0.709	4.90	0.766
0.95	0.565	2.95	0.711	4.95	0.767
1.00	0.572	3.00	0.713	5.00	0.768
1.05	0.579	3.05	0.715	5.05	0.769
1.10	0.585	3.10	0.716	5.10	0.770
1.15	0.591	3.15	0.718	5.15	0.771
1.20	0.597	3.20	0.720	5.20	0.772
1.25	0.603	3.25	0.722	5.25	0.773
1.30	0.608	3.30	0.724	5.30	0.773
1.35	0.613	3.35	0.725	5.35	0.774
1.40	0.618	3.40	0.727	5.40	0.775
1.45	0.622	3.45	0.728	5.45	0.776
1.50	0.627	3.50	0.730	5.50	0.777
1.55	0.631	3.55	0.732	5.55	0.778
1.60	0.635	3.60	0.733	5.60	0.779
1.65	0.639	3.65	0.735	5.65	0.780
1.70	0.643	3.70	0.736	5.70	0.780
1.75	0.647	3.75	0.738	5.75	0.781
1.80	0.650	3.80	0.739	5.80	0.782
1.85	0.654	3.85	0.741	5.85	0.783
1.90	0.657	3.90	0.742	5.90	0.784
1.95	0.661	3.95	0.743	5.95	0.785

Solution (a) After 15 min according to Eq. 3.27 the penetration depth $\delta_{0.01} = 3.6\sqrt{1.5/(900 \cdot 2300) \cdot 15 \cdot 60} = 0.092$ m.
(b) According to Eq. 3.24 $\eta = \frac{0.092}{2 \cdot \sqrt{1.5/(900 \cdot 2300) \cdot 60 \cdot 60}} = 0.90$ and Eq. 3.25 and Fig. 3.9 yields $T = 200$ °C.

3.2.2 Constant Heat Flux: Second Kind of BC

Under some conditions the heat transfer q_s'' to a surface may be assumed constant. According to Sect. 1.1.3 this is a second kind of BC. That may happen, e.g. when the incident radiation to a surface is very high in comparison to the losses by emitted radiation and convection which then can be neglected. Then at a point at a distance x from the surface the temperature is

$$T(x,t) - T_i = \dot{q}_s''\left[\frac{2\sqrt{t}}{\sqrt{\pi}\sqrt{k \cdot \rho \cdot c}} \cdot e^{\left(-\frac{x^2}{4\alpha \cdot t}\right)} - \frac{x}{k}\left(1 - erf\frac{x}{2\sqrt{\alpha \cdot t}}\right)\right] \tag{3.28}$$

where the thermal diffusivity $\alpha = k/(c \cdot \rho)$. At $x = 0$ the surface temperature T_s vs. time becomes

$$T_s - T_i = \frac{2\,\dot{q}_s''\sqrt{t}}{\sqrt{\pi} \cdot \sqrt{k \cdot \rho \cdot c}} \tag{3.29}$$

Thus the time to reach a given temperature rise assuming constant heat flux (for example, time to ignition) at the surface becomes

$$t_{ig} = \frac{\pi \cdot k \cdot \rho \cdot c}{4\,(\dot{q}_s'')^2}\left(T_{ig} - T_i\right)^2 \tag{3.30}$$

where t_{ig} and T_{ig} are the time to ignition and the ignition temperature, respectively.

3.2.3 Constant Gas Temperature: Third Kind of BC

When a surface is exposed to a fluid at a temperature T_g, the heat flux to the surface is

$$\dot{q}_s'' = -k\left(\frac{dT}{dx}\right) = h(T_g - T_s) \tag{3.31}$$

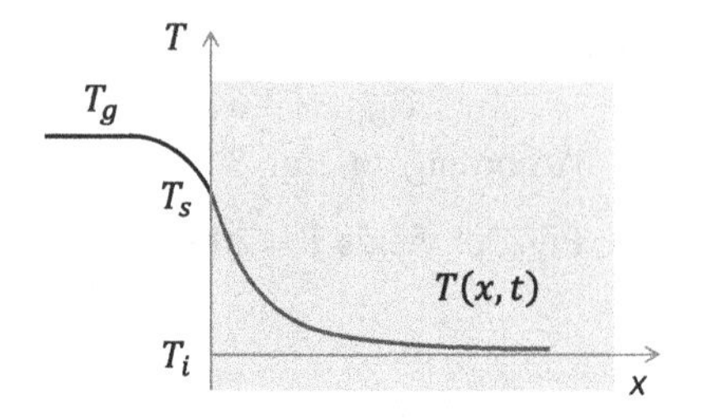

Fig. 3.10 Temperature distribution in a semi-infinite body exposed to a third kind of boundary condition

where h is the heat transfer coefficient and T_s is the surface temperature. This is a third kind of BC according to Sect. 1.1.3. In the case T_g and h are *constant* the temperature distribution may after some time develop as indicated in Fig. 3.10.

Then the relative temperature change at a distance x from the surface can be calculated as

$$\frac{T(x,t)-T_i}{T_g-T_i}=1-erf(X)-e^{\left(\frac{h\cdot x}{k}+\frac{t}{\tau}\right)}\left[1-erf\left(X+\sqrt{\frac{t}{\tau}}\right)\right] \tag{3.32}$$

where the temperature $T(x,t)$ is a function of time and depth and T_i is the initial temperature. The non-dimensional length

$$X=\frac{x}{2\sqrt{\alpha\cdot t}} \tag{3.33}$$

and the time constant for the semi-infinite case is here defined as

$$\tau=\frac{k\cdot\rho\cdot c}{h^2} \tag{3.34}$$

The temperature at the surface is of interest in many fire protection engineering problems such as predictions of time to ignition. The relative temperature change may be obtained from Eq. 3.32 for $x=0$ as

$$\frac{T_s-T_i}{T_g-T_i}=1-e^{\frac{t}{\tau}}\left[1-erf\left(\sqrt{\frac{t}{\tau}}\right)\right] \tag{3.35}$$

or when expressed with the complementary error function as

$$\frac{T_s-T_i}{T_g-T_i}=1-e^{\frac{t}{\tau}}\cdot erfc\left(\sqrt{\frac{t}{\tau}}\right) \tag{3.36}$$

The relative surface temperature rise may also be obtained from the diagram of Fig. 3.11 or from Table 3.3.

Equation 3.33 indicates that the relative surface temperature rise vs. time depends, for a given heat transfer coefficient h, on the material parameter group the *thermal inertia*. The thermal inertia is very important in FSE as it governs how fast a surface reaches among other things ignition temperatures. It varies considerably for many common materials as shown in Table 1.2. Materials of low density ρ have in general also low conductivity k which enhances the differences between materials of various densities. The specific heat capacity varies only relatively little between common materials. More on the influence of thermal inertia on ignition is discussed in Sect. 8.2.

Example 3.4 A 300-mm-thick concrete slab has reinforcement bars at a depth of 30 mm from the bottom surface. The slab is suddenly exposed from below to a fire having a constant temperature of $T_f = 900$ °C. Assume the initial temperature of the slab $T_i = 20$ °C and the thermal conductivity of the concrete, $k = 1.0$ W/(m K), density, $\rho = 2300$ kg/m^3, and the specific heat capacity, $c = 800$ J/(kg K).

(a) What is the surface temperature T_0 of the slab after 10 min of fire exposure? Assume the total heat transfer coefficient due to radiation and convection is constant, $h = 75$ W/m^2 K.
(b) How long does it take until the reinforcement reaches a temperature of 500 °C. Assume in this case that the surface instantly gets the fire temperature, i.e. the heat transfer resistance can be negligible.
(c) Estimate how long it takes until the temperature 300 mm from the bottom of the slab, i.e. at the top surface of the slab, has risen by approximately 10 °C (assuming that the slab is infinitely thick)?

Solution (a) Assume the slab is semi-infinite and apply Eq. 3.35 $\frac{t}{\tau} = h^2 \frac{t}{k\rho c} = 75^2 \frac{10 \cdot 60}{1 \cdot 2300 \cdot 800} = 1.83$ and insert into Eq. 3.35 (or use Fig. 3.11 or Table 3.3) to get $\frac{T_0 - T_i}{T_f - T_i} = 0.65$. Thus $T_0 = 20 + (900 - 20) \cdot 0.65 = 593$ °C.

(b) Apply Eq. 3.22, $\frac{500-20}{900-20} = 0.545 = 1 - erf\left(\frac{x}{2\sqrt{\alpha t}}\right)$ thus $\left(\frac{x}{2\sqrt{\alpha t}}\right) = 0.43$ from both *erf* and *erfc* are shown in Fig. 3.9 as functions of the dimensionless parameter group $x/(2\sqrt{\alpha t})$.

Table 3.2 or Fig. 3.9. Then with $x = 0.03$ m and $\alpha = \frac{1.0}{2300 \cdot 800} = 0.543 \cdot 10^{-6}$ m^2/s the time can be calculated as $t = \frac{0.03^2}{0.543 \cdot 10^{-6} \cdot 2^2 \cdot 0.43^2} = 2241\,\text{s} \approx 37$ min. Apply Eq. 3.27, $\delta_{0.01} = 0.3 = 3.6\sqrt{\alpha t} \Rightarrow t = \frac{0.3^2}{3.6^2 \cdot 0.543 \cdot 10^{-6}} = 12.8 \cdot 10^3$ s $= 3.5$ h.

Example 3.5 The surface temperature of a thick concrete wall with an initial temperature $T_i = 20$ °C rises suddenly to $T_s = 1000$ °C. Assume constant concrete properties according to Table 1.2, i.e. $c = 900$ Ws/(kg K), $\rho = 2300$ kg/m^3 and $k = 1.5$ W/(m K).

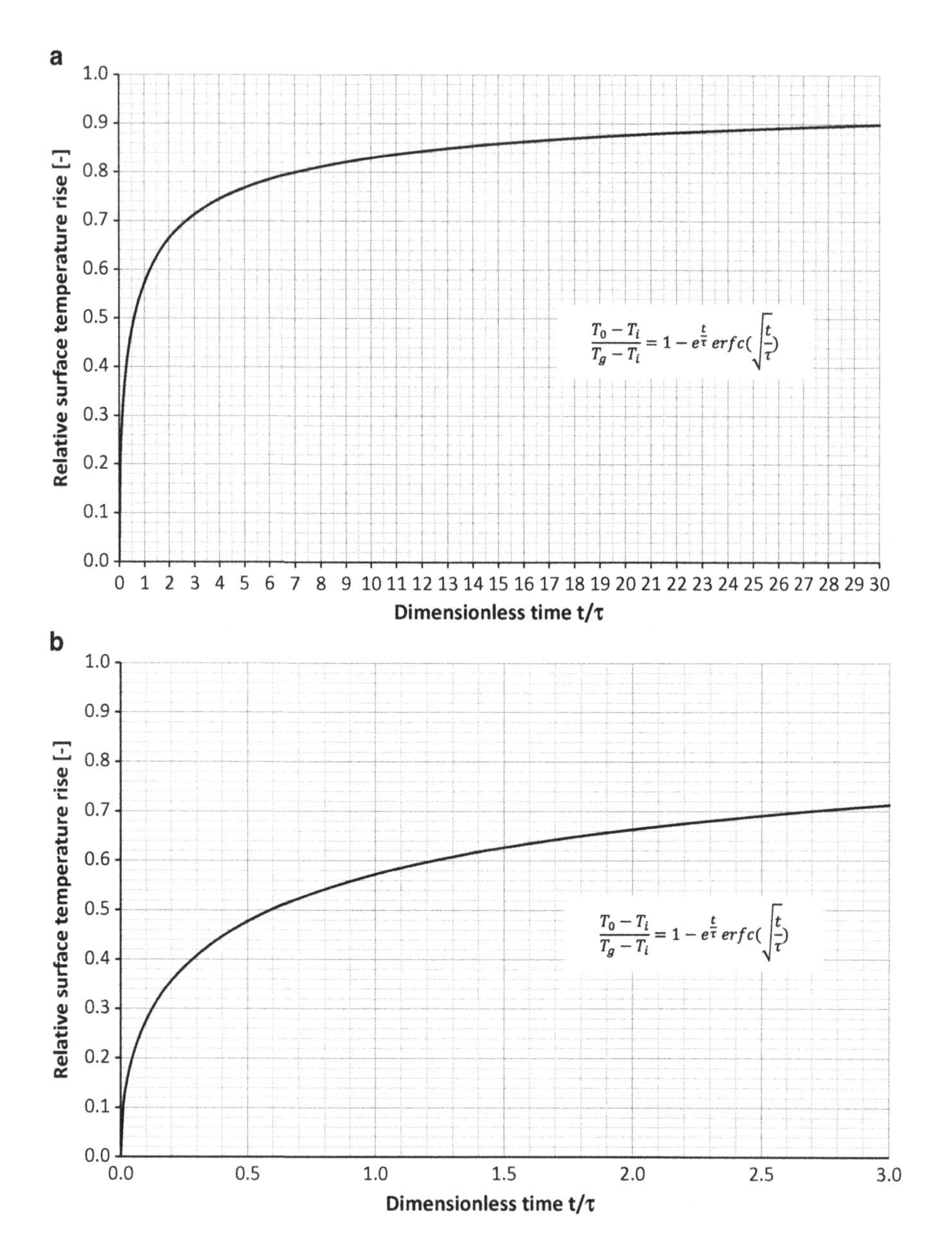

Fig. 3.11 The relative surface temperature change of a semi-infinitely thick body $(T_s - T_i)/(T_g - T_i)$ vs. dimensionless time t/τ according to Eq. 3.35. (**a**) Dimensionless time $t/\tau \leq 30$. (**b**) Dimensionless time $t/\tau \leq 3$

(a) Plot a diagram of the temperature distribution at 30, 60 and 120 min.
(b) What are the temperature penetration depths at 30, 60 and 120 min?
Guidance: *Assume 1 % accuracy*, i.e. $\delta_{0.01} = 3.6\sqrt{\alpha \cdot t}$.

(c) After how long time will the temperature of a reinforcement bar at a depth of 30 mm from the exposed surface start to rise?
Guidance: *Assume 1 % accuracy.*
(d) How thick must the wall be to be considered infinitely thick when calculating the temperature at 30 mm from the exposed surface at 30, 60 and 120 min?
Guidance: *Assume 1 % accuracy and that the temperature change goes to the rear surface and back to the reinforcement bar.*

Solution The temperature diffusion $\alpha = 1.5/(300 \cdot 2300) = 0.821 \cdot 10^{-6}$ m^2/s.

(a) See Fig. 3.12.
(b) Equation 3.27 yields the penetration depth $\delta_{0.01} =$ 138 mm, 196 mm and 277 mm, respectively, for 30, 60 and 120 min.
(c) Equation 3.27 yields $t = \left(\frac{0.03}{3.6}\right)^2 / \left(0.821 \cdot 10^{-6}\right) = 84.5$ s.
(d) The distance x + (x − 0.03) must be longer than the penetration depths (see arrows in Fig. 3.13). Thus the wall thickness $x \geq 0.5 \cdot \delta_{0.01} + 0.03$ which yields the thicknesses 99, 128 and 168 mm, respectively, for 30, 60 and 120 min.

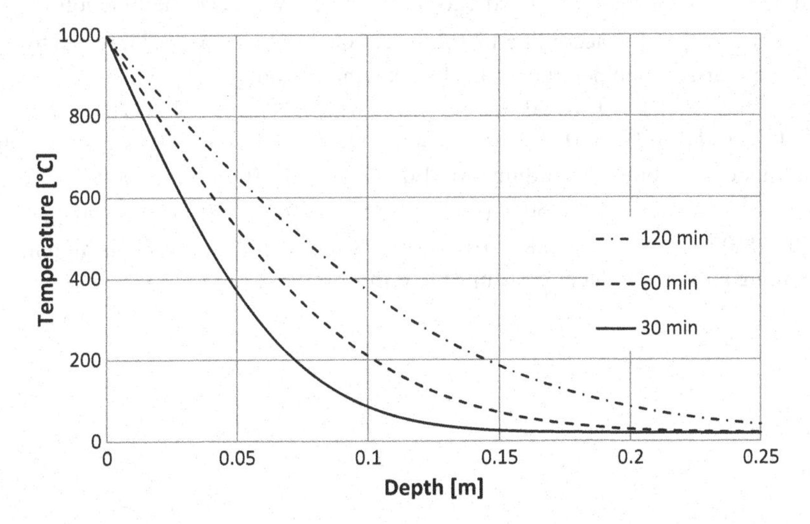

Fig. 3.12 Temperature distributions at various times. Example 3.5

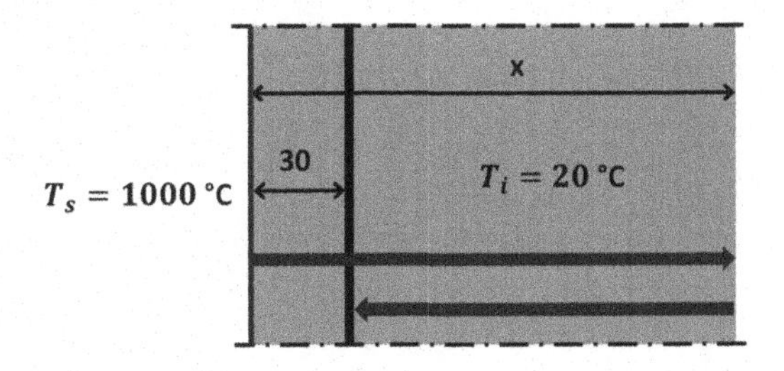

Fig. 3.13 Estimation whether a wall can be considered infinitely thick when calculating the temperature of a reinforcement bar

Example 3.6 A very thick concrete wall is penetrated by a steel beam. It is suddenly exposed to high temperature on one side. Estimate roughly how long it takes before the temperature rise is felt on the unexposed side.

(a) On the concrete surface away from the beam.
(b) On the steel beam surface.

Material properties according to Table 1.2.

Solution The thermal diffusivity of concrete is $0.82 \cdot 10^{-6}$ m^2/s and for mild steel $12.7 \cdot 10^{-6}$. Then Eq. 3.26 yields $t = \frac{1}{\alpha} \cdot \left(\frac{\delta_{0.05}}{2.8}\right)^2$, i.e. $t = \frac{1}{0.82 \cdot 10-6} \cdot \left(\frac{0.02}{2.8}\right)^2 = 62$ s for the concrete surface and only 4 s for the steel surface.

Example 3.7 The surface of a thick pine wood panel with an initial temperature $T_i = 20$ °C is suddenly exposed to hot gases with a temperature of $T_g = 600$ °C. Assume a constant heat transfer coefficient $h = 50$ W/(m^2 K). What is the surface temperature T_s and the heat flux $\dot{q}_s''$ at time $t = 0$, 30 and 120 s. Assume thermal properties of pine according to Table 1.2.

Solution According to Eq. 3.34 $\tau = (k \cdot \rho \cdot c)/h^2 = 0.196 \cdot 10^6/50^2 = 78.4$ s. Then $t/\tau = 0/78.4$, 30/78.4 and 120/78.4, respectively, and the function $\big(1 - \exp(t/\tau) \cdot \text{erfc}\big(\sqrt{\frac{t}{\tau}}\big)$ is according to Table 3.3 equal to 0, 55 and 0.73, respectively, and the surface temperature can be obtained from Eq. 3.35 as $T_s(0) = 20 + (600 - 20) \cdot 0 = 20$ °C, $T_s(30) = 20 + (600 - 20) \cdot 0.44 = 275$ °C and $T_s(120) = 20 + (600 - 20) \cdot 0.63 = 385$ °C. The corresponding heat fluxes become according to Eq. 3.31 $\dot{q}_s''(0) = 50 \cdot (600 - 20) = 29 \cdot 10^3$ W/m^2, $\dot{q}_s''(30) = 50 \cdot (600 - 275) = 16.3 \cdot 10^3$ W/m^2 and $\dot{q}_s''(120) = 50 \cdot (600 - 385) = 10.8 \cdot 10^3$ W/m^2. Notice that the heat flux is high in the beginning and then reduced by almost two thirds after 120 s.

Chapter 4
Boundary Conditions in Fire Protection Engineering

A summary of the three kinds of boundary conditions as outlined in Sect. 1.1.3 is shown in Table 4.1. The third kind of BC sometimes called natural BC is by far the most important and common boundary condition in fire protection engineering, while the first and second kinds of BCs can rarely be specified. The third kind of BC may be divided into three subgroups, (a), (b) and (c). The subgroup (b) and (c) are particularly suitable for fire engineering applications. Subgroup (a) is applied when the heat transfer coefficient may be assumed constant as assumed in Chaps. 2 and 3. T_g is then the surrounding gas temperature. In fire protection engineering it is, however, generally not accurate enough to assume a constant heat transfer coefficient as in particular heat transfer by radiation is highly non-linear, i.e. the heat transfer coefficient varies with the surface temperature. Therefore the subgroups (3b) and (3c) are the most commonly applied. They consist of a radiation term and a convection term with the corresponding emissivity ε and convection heat transfer coefficient h, respectively. The subgroup (3b) presupposes a uniform temperature T_f, i.e. the radiation temperature and the gas temperature are equal. This is assumed, for example, when applying time–temperature design curves according to standards such as ISO 834 or EN 1363-1 for evaluating the fire resistance of structures, see Chap. 12. The subgroup (3c) is a more general version of (3b) as it allows for different gas T_g and radiation T_r temperatures, so-called *mixed boundary conditions*. Alternatively $\sigma \cdot T_r^4$ may be replaced by an equivalent specified incident radiation $\dot{q}''_{inc}$ according to the identity $\dot{q}''_{inc} \equiv \sigma \cdot T_r^4$ (Eq. 1.17). As shown in Sect. 4.4 all boundary conditions of subgroup 3 may be written as type 3a. That means momentarily a single effective temperature named the *adiabatic surface temperature* (AST) with a value between the radiation and gas temperatures as well as a corresponding total heat transfer coefficient can always be defined, see Sect. 4.4.

All the specified boundary conditions given in Table 4.1 may vary with time. In most calculations the emissivity and the convection heat transfer coefficient are, however, assumed constant while the radiation and gas temperatures may vary according to standard, measured or calculated values, see Chaps. 10–12.

U. Wickström, *Temperature Calculation in Fire Safety Engineering*,
DOI 10.1007/978-3-319-30172-3_4

Table 4.1 Summary of the three kinds of boundary conditions. The third kind is divided into three subgroups relevant in FSE

No	Kind of boundary condition	Formula
1	Prescribed surface temperature	$T_{x=x_0} = T_s$
2	Prescribed surface heat flux	$-k\frac{\partial T}{\partial x}\Big\vert_{x=0} = \dot{q}''_s$
(3a)	Natural boundary condition (prescribed convection)	$-k\frac{\partial T}{\partial x}\Big\vert_{x=0} = h(T_g - T_s)$
(3b)	Natural boundary condition (prescribed convection and radiation, equal radiation and gas temperatures)	$-k\frac{\partial T}{\partial x}\Big\vert_{x=0} = \varepsilon\sigma\left(T_f^4 - T_s^4\right) + h_c(T_f - T_s)$
(3c)	Natural boundary condition (prescribed convection and radiation conditions, different radiation and gas temperatures)	$-k\frac{\partial T}{\partial x}\Big\vert_{x=0} = \varepsilon \cdot \sigma(T_r^4 - T_s^4) + h_c(T_g - T_s)$ or $-k\frac{\partial T}{\partial x}\Big\vert_{x=0} = \varepsilon(\dot{q}''_{inc} - \sigma T_s^4) + h_c(T_g - T_s)$ or $-k\frac{\partial T}{\partial x}\Big\vert_{x=0} = h_{AST,tot}(T_{AST} - T_s)$

4.1 Radiation and Incident Radiation Temperature

The *black body radiation temperature* T_r was introduced in Sect. 1.1.3 by the identity

$$\dot{q}''_{inc} \equiv \sigma \cdot T_r^4 \tag{4.1}$$

or reversely

$$T_r \equiv \sqrt[4]{\frac{\dot{q}''_{inc}}{\sigma}} \tag{4.2}$$

A more adequate term would be *incident black body radiation temperature* as $\dot{q}''_{inc}$ depends on direction. By definition T_r is the temperature of a surface in equilibrium with the incident radiation, i.e. the absorbed heat by radiation is equal to the emitted heat.

Figure 4.1 shows the relation between the incident radiation $\dot{q}''_{inc}$ and the radiation temperature T_r as defined by Eq. 4.1. T_r may be given in Kelvin as in Eq. 4.1 (lower curve) and in °C (upper curve). The temperature shift between the two temperature scales is 273.15 K, i.e. [temperature in Kelvin] = [temperature in °C + 273.15]. In Table 4.2 the relations between $\dot{q}''_{rad}$ and T_r are given at selected levels. Thus, for example, an incident radiant flux of 20 kW/m^2 corresponds to a black body radiation temperature of 771 K = 498 °C, and a radiation temperature of 1000 °C = 1273 K corresponds to an incident radiant flux of 148.9 kW/m^2.

The net heat flux by radiation to a surface is according to Eq. 1.16

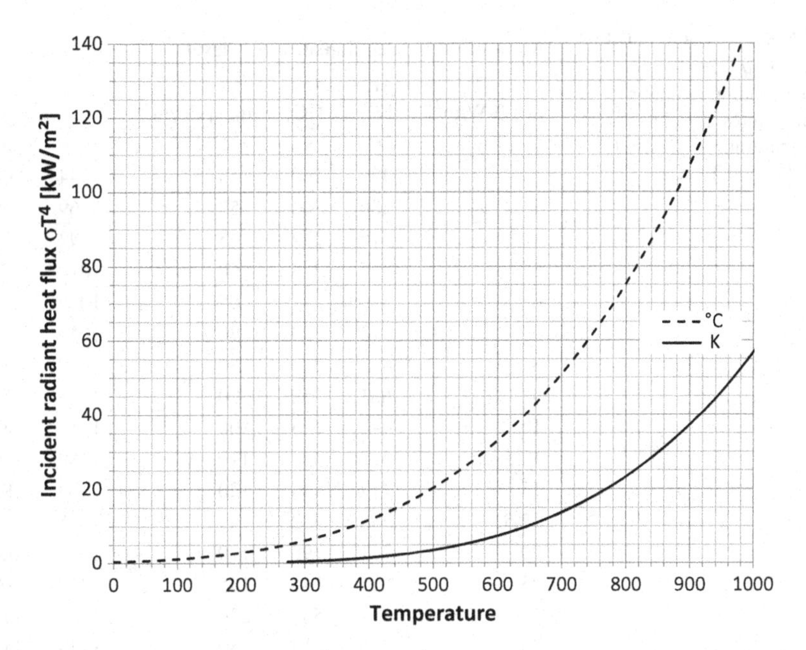

Fig. 4.1 Incident radiation heat flux $\dot{q}''_{inc}$ vs. incident radiation temperature T_r. The *upper curve* refers to temperature in °C and the lower to Kelvin as in Eq. 4.1

$$\dot{q}''_{rad} = \varepsilon \cdot \sigma\left(T_r^4 - T_s^4\right) \tag{4.3}$$

Alternatively it can be "linearized" and written as

$$\dot{q}''_{rad} = h_r(T_r - T_s) \tag{4.4}$$

where the *radiation heat transfer coefficient* h_r is obtained by developing the parentheses of Eq. 1.16 according to the conjugate rule:

$$h_r = \varepsilon \cdot \sigma\left(T_r^2 + T_s^2\right) \cdot (T_r + T_s) \tag{4.5}$$

As shown in Fig. 4.2 the radiation heat transfer coefficient h_r varies significantly depending on the radiation and surface temperatures. At room temperature it is less than 5 W/(m^2 K) while it is between 150 and 400 W/(m^2 K) or even more at temperature levels relevant in fire scenarios.

In many cases T_r and T_s are close and may be assumed equal. Then h_r can be approximated as

$$h_r \approx 4\,\varepsilon \cdot \sigma \cdot T_r^3 \tag{4.6}$$

and the radiation heat transfer coefficient becomes then depending on the incident radiation

Table 4.2 Incident radiation heat flux $\dot{q}''_{inc}$ and corresponding radiation temperatures, absolute T_r [K] according to Eqs. 4.1 and 4.2, and T [°C]

(a) Selected incident radiation $\dot{q}''_{inc}$ levels

	T_r	T		T_r	T
q_{inc} [kW/m²]	[K]	[°C]	q_{inc} [kW/m²]	[K]	[°C]
1	364	91	40	916	643
2	433	160	45	944	671
3	480	206	50	969	696
4	515	242	55	992	719
5	545	272	60	1014	741
6	570	297	65	1035	762
7	593	320	70	1054	781
8	613	340	80	1090	817
9	631	358	90	1122	849
10	648	375	100	1152	879
12.5	685	412	110	1180	907
15	717	444	120	1206	933
17.5	745	472	145	1265	991
20	771	498	170	1316	1043
25	815	542	195	1362	1089
30	853	580	220	1403	1130
35	886	613	250	1449	1176

(b) Selected radiation temperature $(T_r - 273)$ °C levels

	T_r	q_{inc}		T_r	q_{inc}
T [°C]	[K]	[kW/m²]	T [°C]	[K]	[kW/m²]
100	373	1.10	550	823	26.01
120	393	1.35	700	973	50.82
140	413	1.65	750	1023	62.10
160	433	1.99	800	1073	75.16
180	453	2.39	850	1123	90.18
200	473	2.84	900	1173	107.34
225	498	3.49	950	1223	126.85
250	523	4.24	1000	1273	148.90
300	573	6.11	1050	1323	173.71
325	598	7.25	1100	1373	201.50
350	623	8.54	1150	1423	232.49
375	648	10.00	1200	1473	266.93
400	673	11.63	1250	1523	305.06
425	698	13.46	1300	1573	347.13
450	723	15.49	1350	1623	393.42
475	748	17.75	1400	1673	444.19
500	773	20.24	1450	1723	499.72

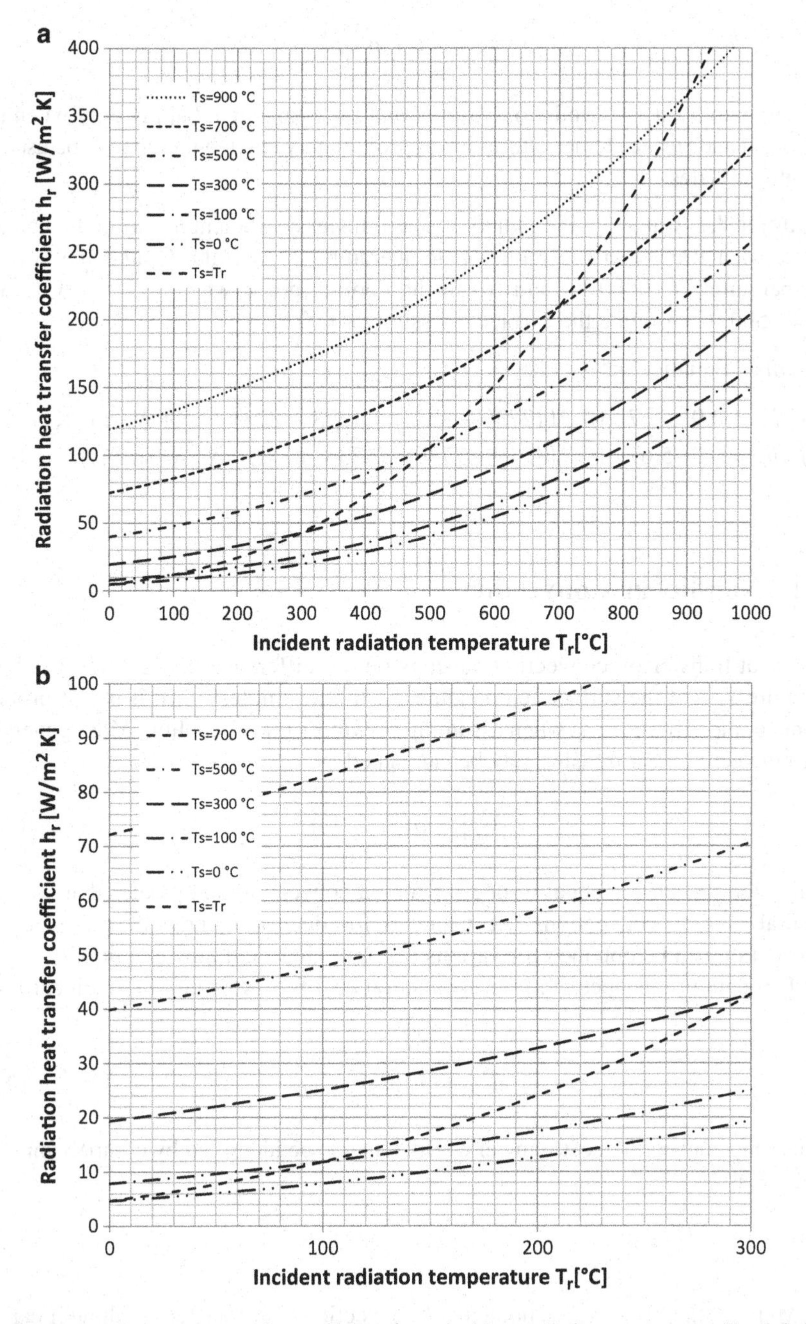

Fig. 4.2 The radiation heat transfer coefficient h_r vs. incident radiation temperature (Eq. 4.5) for various surface temperatures T_s assuming $\varepsilon = 1$. (**a**) $T_r \leq 1000\ °C$. (**b**) Enlargement, $T_r \leq 300\ °C$

$$h_r \approx 4\,\varepsilon\,\sqrt[4]{\sigma \cdot \dot{q}^{3}_{inc}} \tag{4.7}$$

by combining Eqs. 4.1 and 4.6. Observe that Eqs. 4.6 and 4.7 apply only when the radiation and surface temperatures are approximately equal but they may be used as rough estimates.

Example 4.1 Calculate the radiation heat transfer coefficient h_r, (a) in the heating phase of a fire when the radiation temperature $T_r = 1000\ °C$ and the surface temperature $T_s = 600\ °C$ and (b) in the cooling phase when $T_r = 200\ °C$ and $T_s = 500\ °C$. Assume the surface emissivity $\varepsilon = 0.9$.

Solution Equation 4.5 yields:

(a) $h_r = 0.9 \cdot 5.67 \cdot 10^{-8}\left(1273^2 + 873^2\right) \cdot (1273 + 873) = 261\ \mathrm{W/(m^2\ K)}$.
(b) $h_r = 0.9 \cdot 5.67 \cdot 10^{-8}\left(473^2 + 773^2\right) \cdot (473 + 873) = 52\ \mathrm{W/(m^2\ K)}$.

4.2 Non-linear Convection

The heat transfer by convection depends on the difference between the gas temperature T_g and the surface temperature T_s. In the simplest form it is just proportional to the difference as when assuming Newton's law of cooling. More generally the convection heat transfer may be calculated as

$$\dot{q}''_{con} = \beta\left(T_g - T_s\right)^{\gamma} \tag{4.8}$$

where the power γ is equal to one for forced convection and greater than one for natural or free convection. See Chap. 6 for details on how heat transfer by convection can be obtained for various configurations and flow conditions.

Throughout this document the convection heat flux is written in the linear form as

$$\dot{q}''_{con} = h_c\left(T_g - T_s\right) \tag{4.9}$$

where the convection heat transfer coefficient can be identified by comparison with Eq. 4.8 as

$$h_c = \beta\left(T_g - T_s\right)^{(\gamma-1)} \tag{4.10}$$

More on the physical phenomena of convection heat transfer and how it can be estimated is given in Chap. 6.

4.3 Mixed Boundary Conditions

The total heat flux to a surface is the sum of the radiation and the convection contributions and may according to Eqs. 4.3 and 4.9 be written as

$$\dot{q}''_{tot} = \varepsilon \left(\dot{q}''_{inc} - \sigma \cdot T_s^4 \right) + h_c \left(T_g - T_s \right) \quad (4.11)$$

or by expressing the incident radiant flux by the radiation temperature as defined by Eq. 4.1

$$\dot{q}''_{tot} = \varepsilon \cdot \sigma \left(T_r^4 - T_s^4 \right) + h_c \left(T_g - T_s \right) \quad (4.12)$$

This equation contains two boundary temperatures, the radiation temperature and the gas or convection temperature. It may then be called *mixed boundary* conditions.

An electric circuit analogy of a mixed boundary condition with two temperatures and two corresponding heat transfer resistances is shown in Fig. 4.3.

If the radiation heat transfer coefficient h_r as defined by Eqs. 4.5 or 4.6 is used we can get

$$\dot{q}''_{tot} = h_r (T_r - T_s) + h_c \left(T_g - T_s \right) \quad (4.13)$$

or in terms of thermal resistances

$$\dot{q}''_{tot} = \left[(T_r - T_s)/R_r + \left(T_g - T_s \right)/R_c \right] \quad (4.14)$$

where the radiation heat transfer resistance (cf. Eqs. 2.14 and 4.5)

$$R_r = \frac{1}{h_r} = \frac{1}{\varepsilon \cdot \sigma \left(T_r^2 + T_s^2 \right) \cdot (T_r + T_s)} \quad (4.15)$$

and the corresponding convection heat transfer resistance

$$R_c = \frac{1}{h_c} \quad (4.16)$$

In most fire resistance cases and calculation standards such as EN 1991-1-2 (Eurocode 1) dealing with exposure to post-flashover room fires, the radiation temperature and the gas temperature are assumed equal to a fire temperature $T_f = T_r = T_g$. Then the total heat transfer becomes

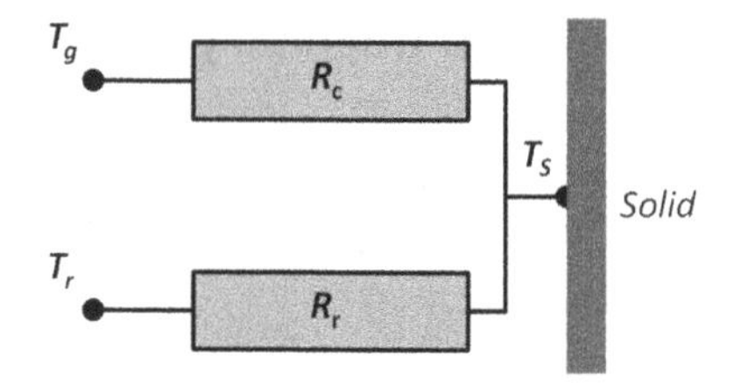

Fig. 4.3 Electric circuit analogy of mixed boundary conditions according to Eq. 4.14

$$\dot{q}''_{tot} = \varepsilon \cdot \sigma \left(T_f^4 - T_s^4 \right) + h_c (T_f - T_s) \tag{4.17}$$

This equation may also be written as

$$\dot{q}''_{tot} = h_{tot} (T_f - T_s) = \frac{(T_f - T_s)}{R_{tot}} \tag{4.18}$$

where

$$h_{tot} = h_r + h_c = \varepsilon \cdot \sigma \left(T_f^2 + T_s^2 \right) \cdot (T_f + T_s) + h_c \tag{4.19}$$

and

$$R_{tot} = \frac{1}{h_{tot}} \tag{4.20}$$

Example 4.2 A surface in air at ambient temperature of $T_\infty = 25$ °C is exposed to radiation from a thick flame at a temperature of 800 °C. Assume the surface emissivity $\varepsilon = 1$ and the convection heat transfer coefficient $h_c = 50$ W/m^2 K?

(a) What is the radiation heat transfer coefficient h_r if the surface temperature T_s is 600 °C.
(b) Use the calculated h_r to calculate the total heat transfer $\dot{q}''_{tot}$ to the surface.

Solution
(a) Equation 4.5 yields $h_r = 211$ W/m^2.
(b) Equation 4.13 yields $\dot{q}''_{tot} = [211 \cdot (800 - 600) + 50 \cdot (25 - 600)]$ W/m^2 $= 13{,}500$ W/m^2.

4.4 Adiabatic Surface Temperature

Thermal exposure of a surface depends according to Eqs. 4.11 and 4.12 on *two* independent parameters T_r (or $\dot{q}''_{inc}$) and T_g and can then in principle not be expressed by one single parameter. The radiation and gas temperatures are in general *not* equal. The radiation temperature may be either higher or lower than the adjacent gas temperature. Equations 4.11 or 4.12 can always be applied. Alternatively, however, the heat transfer may be written with one parameter only, given the relation between the emissivity and the convective heat transfer coefficient h_c/ε is known. Then an artificial effective temperature denoted AST T_{AST} can replace T_r and T_g. A very important advantage of introducing AST is that it can be measured also under harsh fire conditions with the robust so-called Plate Thermometers as described in Sect. 9.3, and it can be obtained from numerical calculations with fire modelling codes such as FDS (Fire Dynamic Simulator) [10].

The AST depends on position as well as on direction. For example, at a point outside a fire as illustrated in Fig. 4.4 the highest incident radiation is in the direction A from the fire while from other directions it is less. Therefore in this case $T^A_{AST} > T^B_{AST}$. In general at any point in space six different incident fluxes can be identified and thereby six different ASTs, but only one gas temperature T_g. However, in most cases it is obvious that only one direction is of interest, namely perpendicular to an exposed surface.

By definition T_{AST} is the temperature of a surface which cannot absorb any heat, i.e.

$$\varepsilon\left(\dot{q}''_{inc} - \sigma \cdot T^4_{AST}\right) + h_c\left(T_g - T_{AST}\right) = 0 \tag{4.21}$$

and with the relation between $\dot{q}''_{inc}$ and T_r according to Eq. 4.1

$$\varepsilon \cdot \sigma\left(T^4_r - T^4_{AST}\right) + h_c\left(T_g - T_{AST}\right) = 0 \tag{4.22}$$

T_{AST} is a weighted average value of the radiation temperature T_r and the gas temperatures T_g depending on the surface emissivity ε and the convection heat transfer coefficient h_c. Thus it is a function of T_r, T_g and the parameter ratio h_c/ε, but *independent* of the surface temperature T_s of the exposed body. From Eq. 4.22 it is evident that T_{AST} has a value between T_r and T_g as being illustrated by Fig. 4.5. The larger values of h_c/ε, the closer T_{AST} will be to T_g, and vice versa the smaller values of h_c/ε the closer the value of T_{AST} will be to T_r. In other words, when the heat transfer by convection is dominating T_{AST} is near the gas temperature and when the radiation is dominating it is closer to the radiation temperature.

The AST may be derived from Eq. 4.22 as

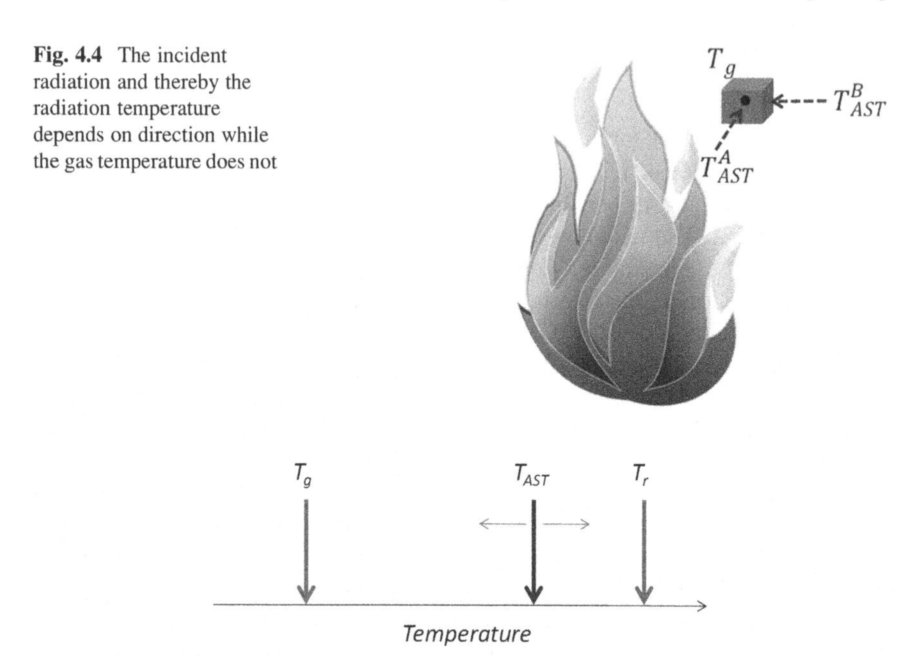

Fig. 4.4 The incident radiation and thereby the radiation temperature depends on direction while the gas temperature does not

Fig. 4.5 The adiabatic surface temperature T_{AST} is always between the radiation temperature T_r and the gas temperature T_g. The higher value of the parameter ratio h_c/ε, the closer T_{AST} will be to T_g, and vice versa

$$T_{AST} = \frac{h_r\,T_r + h_c\,T_g}{h_r + h_c} \tag{4.23}$$

The equation is, however, implicit as h_r depends on $\boldsymbol{T_{AST}}$.

By combining the general heat transfer equations 4.12 and 4.22 the total heat transfer to a surface may alternatively be calculated as

$$\dot{q}''_{tot} = \varepsilon \cdot \sigma\left(T_{AST}^4 - T_s^4\right) + h_c(T_{AST} - T_s) \tag{4.24}$$

Instead of two temperatures, T_r and T_g, the fire temperature level is now in Eq. 4.24 expressed only by one temperature T_{AST}. This may have computational advantages but most important T_{AST} can be measured even at very harsh thermal conditions with so-called Plate Thermometers, see Sect. 9.3.

Figure 4.6 illustrates how the two exposure boundary temperatures T_r and T_g are combined into one effective exposure boundary temperature, namely the AST T_{AST}. This alteration does not introduce any further approximations of the heat transfer conditions.

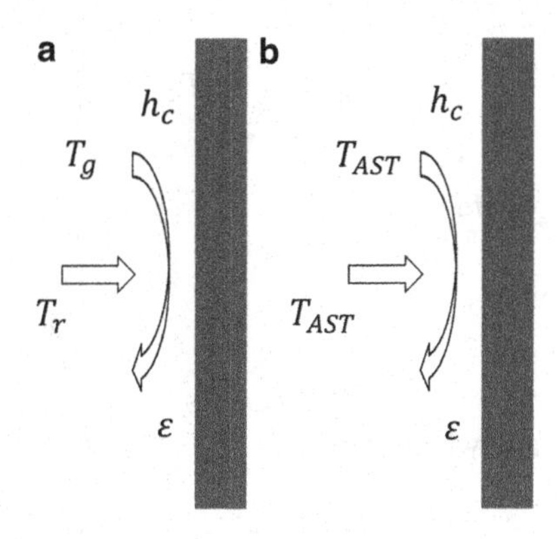

Fig. 4.6 The heat exposure of a surface expressed in terms of (**a**) the radiation temperature T_r and the gas temperature T_g or alternatively in terms of (**b**) the adiabatic surface temperature T_{AST}

4.4.1 Calculation of Adiabatic Surface Temperature and Incident Radiation

When the incident radiation flux $\dot{q}''_{inc}$ (or the equivalent T_r) and the corresponding ε and h_c are known, the AST T_{AST} can be obtained by solving the fourth degree equation according to either Eq. 4.21 or Eq. 4.22. Below two iteration schemes and one exact method are mentioned.

In many cases when the radiation is dominating T_{AST} can be obtained by the iteration procedure

$$T_{AST}^{i+1} = \sqrt[4]{T_r^4 + \frac{h_c}{\varepsilon \cdot \sigma}\left(T_g - T_{AST}^i\right)} \qquad (4.25)$$

where the suffix i and i + 1 denotes the iteration number. By starting the iteration with $T_{AST}^1 = T_r$ the result converges generally within a few iteration steps. Otherwise especially when the convection is dominating a Newton–Raphson iteration scheme may be needed.

An exact solution has been presented by Malendowski (personal communication). Then after elementary algebraic operations, Eq. 4.21 can be written as:

$$\varepsilon \cdot \sigma \cdot T_{AST}^4 + h_c T_{AST} + \left(-\varepsilon \cdot \dot{q}''_{inc} - h_c T_g\right) = 0 \qquad (4.26)$$

which is the fourth order polynomial equation with $\boldsymbol{T_{AST}}$ as the variable. It may be written in the form:

$$a \cdot T_{AST}^4 + b \cdot T_{AST} + c = 0 \qquad (4.27)$$

where the coefficients of the polynomial can be identified as: $a = \varepsilon \cdot \sigma$, $b = h_c$ and $c = -\left(\varepsilon \cdot \dot{q}''_{inc} + h_c T_g\right)$. Eq. 4.27 has generally four roots but the only physical can in the actual case be written as:

$$T_{AST} = \frac{1}{2} \cdot \left(-M + \sqrt{\frac{2b}{aM} - M^2} \right) \tag{4.28}$$

where

$$M = \sqrt{\frac{\beta}{\alpha} + \frac{\alpha}{\gamma}} \tag{4.29}$$

and where

$$\alpha = \sqrt[3]{\sqrt{3} \cdot \sqrt{27a^2b^4 - 256a^3c^3} + 9a \cdot b^2} \quad \beta = 4\sqrt[3]{\frac{2}{3}} \cdot c \quad \gamma = \sqrt[3]{18} \cdot a \tag{4.30}$$

Thus by inserting the parameters of Eq. 4.30 into Eq. 4.29 and then into Eq. 4.28 the solution may be expressed in an exact closed form.

Examples of AST vs. incident radiation temperature for various gas temperature levels and relations between surface emissivity and convection heat transfer coefficient are shown in the graphs of Fig. 4.7a–d.

When T_{AST} is obtained, e.g. by measurements with PTs, the incident radiation $\dot{q}''_{inc}$ can be derived from Eq. 4.21 as

$$\dot{q}''_{inc} = \sigma \cdot T_{AST}^4 - \frac{h_c}{\varepsilon}\left(T_g - T_{AST}\right) \tag{4.31}$$

The accuracy of Eq. 4.31 depends very much on the accuracy of the parameter ratio h_c/ε. However, in most cases at elevated temperature the second term on the right-hand side is small and therefore the accuracy can be relatively high in comparison to alternative instruments available in practice. See also Sect. 9.3.2 how the so-called plate thermometer can be used for indirectly measuring incident radiant heat flux by measuring T_{AST} and then applying Eq. 4.31.

4.4.2 An Electric Circuit Analogy of the AST Boundary Condition

The radiation term of Eq. 4.24 may be developed in a similar way as shown by Eq. 4.5. Then the adiabatic radiation heat transfer coefficient h_r^{AST} may be introduced, and the heat flux can be written as

$$\dot{q}''_{tot} = h_r^{AST}\left(T_{AST} - T_s\right) + h_c^{AST}\left(T_{AST} - T_s\right) \tag{4.32}$$

or in terms of heat transfer resistance according to Eq. 4.20, the radiation and convection heat transfer resistances over an area A are the inverses of the heat transfer coefficients, h_r^{AST} and h_c^{AST},

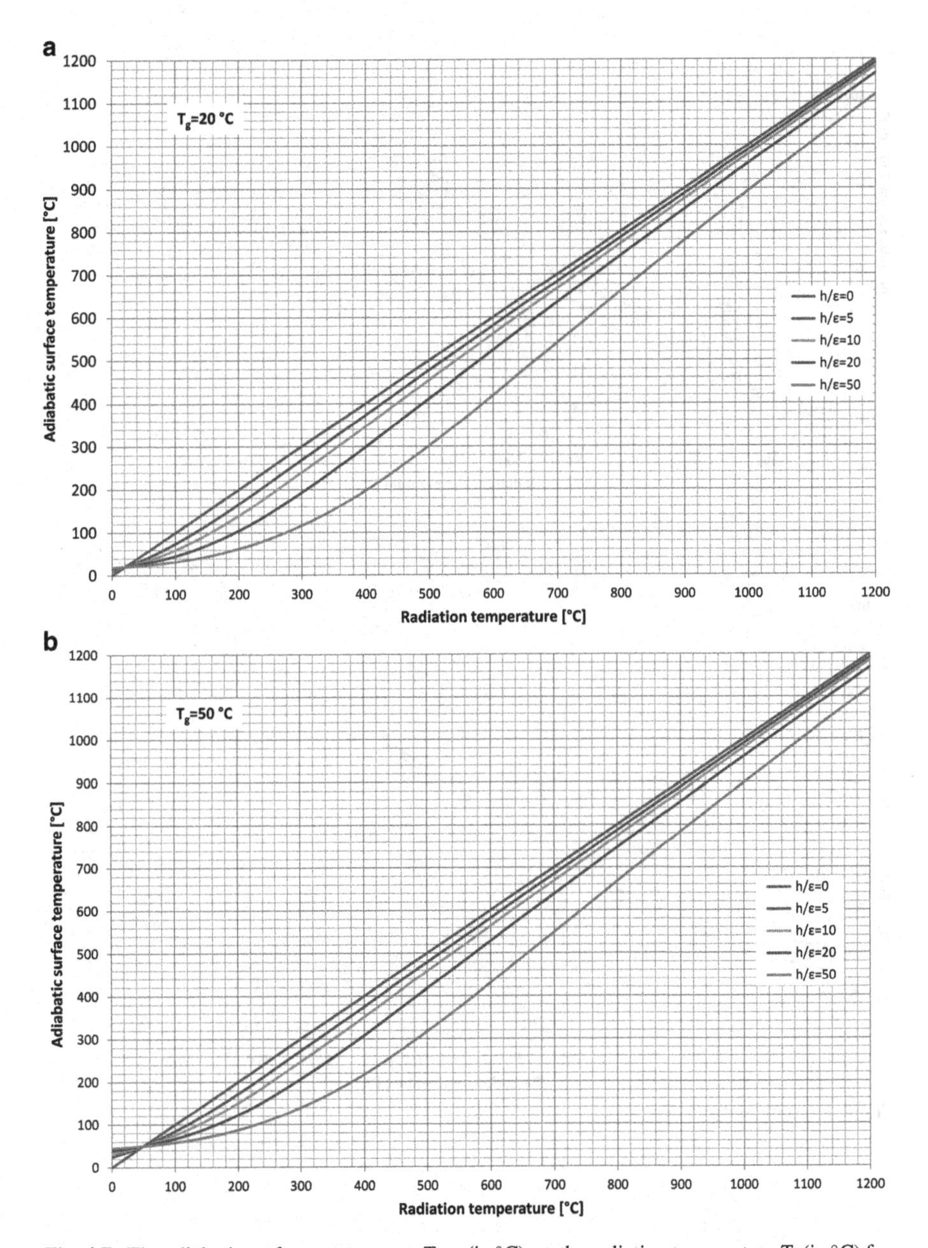

Fig. 4.7 The adiabatic surface temperature T_{AST} (in °C) vs. the radiation temperature T_r (in °C) for various ratio between the convection heat transfer coefficient and the emissivity as defined by Eq. 4.22. Diagrams for gas temperatures $T_g = 20,\ 50,\ 100$ and 500 °C, respectively

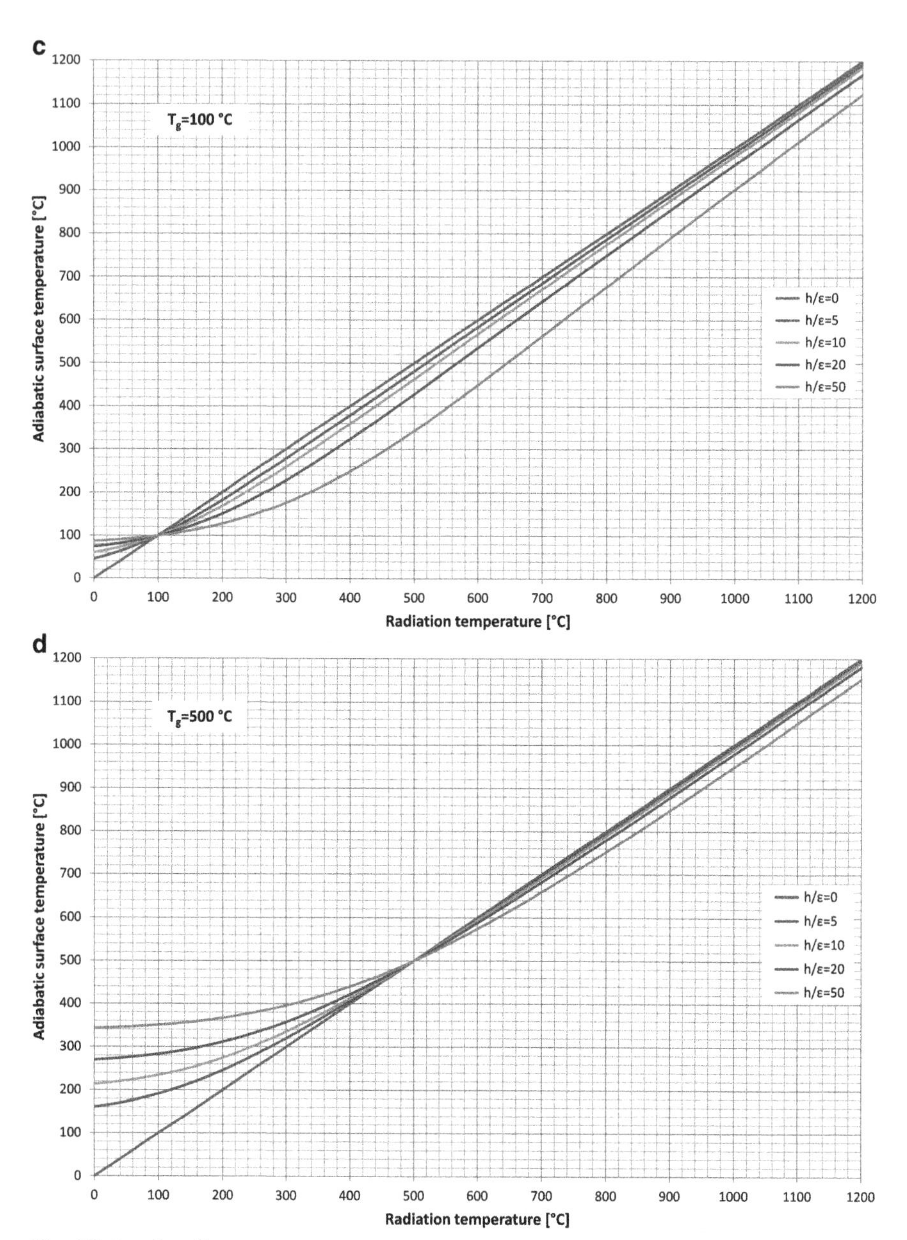

Fig. 4.7 (continued)

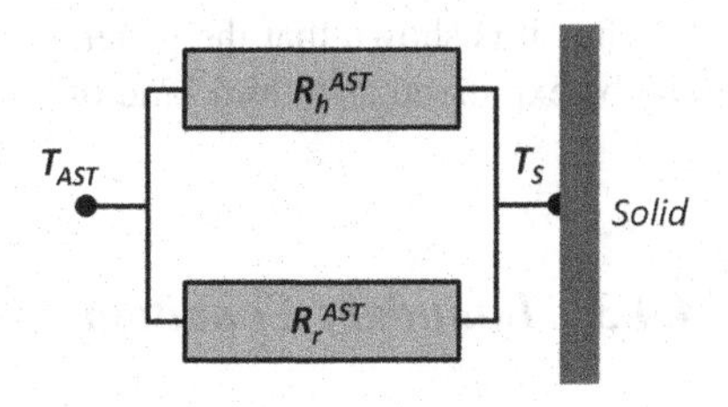

Fig. 4.8 An electric circuit analogy of mixed boundary conditions at a surface of a solid according to Eq. 4.33

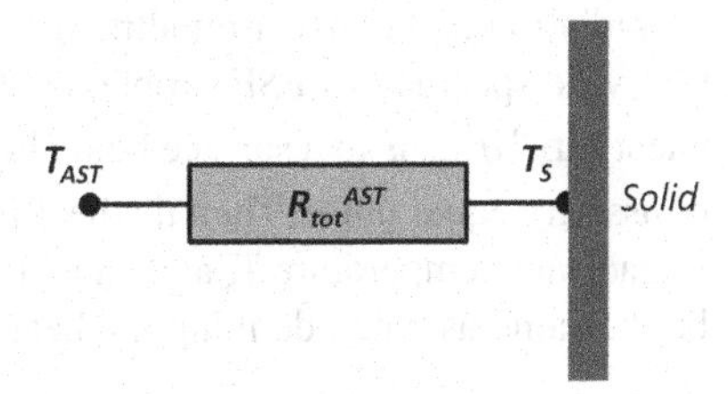

Fig. 4.9 An electric circuit analogy of mixed boundary conditions at a surface of a solid according to Eq. 4.34

$$\dot{q}''_{tot} = \frac{1}{R_r^{AST}} \cdot (T_{AST} - T_s) + \frac{1}{R_c^{AST}} \cdot (T_{AST} - T_s) \tag{4.33}$$

The electric circuit analogy is shown in Fig. 4.8. A total adiabatic heat transfer coefficient h_{tot}^{AST} and a corresponding total adiabatic heat transfer resistance R_{tot}^{AST} can also be defined. Then the total heat transfer to a surface may be written as

$$\dot{q}''_{tot} = h_{tot}^{AST}(T_{AST} - T_s) = (T_{AST} - T_s)/R_{tot}^{AST} \tag{4.34}$$

where the total adiabatic heat transfer coefficient becomes

$$h_{tot}^{AST} = h_r^{AST} + h_c^{AST} \tag{4.35}$$

The adiabatic radiation heat transfer coefficient then becomes

$$h_r^{AST} = \varepsilon \cdot \sigma\left(T_{AST}^2 + T_s^2\right) \cdot (T_{AST} + T_s) \tag{4.36}$$

and the total adiabatic heat transfer resistance becomes

$$R_{tot}^{AST} = \frac{1}{h_r^{AST} + h_c^{AST}} \tag{4.37}$$

The convection heat transfer coefficient remains the same as it is here assumed independent of the exposure temperature, i.e.

$$h_c^{AST} = h_c \tag{4.38}$$

A corresponding electric circuit analogy of a mixed boundary condition is shown in Fig. 4.9.

Thus it is shown that the general fire boundary condition according to Eq. 4.12 can be expressed as a third kind of boundary condition, see Table 4.1.

4.4.3 Boundary Condition Expressed as "Heat Flux"

The thermal exposure conditions are often in FSE literature specified as "heat flux" by radiation and convection although boundary conditions of the second kind can rarely be specified in FSE problems. It is, however, implicitly understood that the "heat flux" $\dot{q}''_{\text{flux}}$ is to a surface being kept at ambient temperature T_∞ and having an emissivity equal unity. Then the heat flux $\dot{q}''_{tot}$ by radiation and convection to a real surface at a temperature T_s and an assumed convective heat transfer coefficient h to be the same as when defining the heat flux becomes:

$$\dot{q}''_{tot} = \dot{q}''_{\text{flux}} - \sigma\left(T_s^4 - T_\infty^4\right) - h(T_s - T_\infty) \tag{4.39}$$

This is now a boundary condition of the third kind as the heat flux to the surface depends on the receiving surface temperature T_s. By comparing the heat fluxes as expressed by Eqs. 4.39 and 4.31, it can be shown that for given values of the heat transfer parameters ε and h there is an unambiguous relation between the two artificial boundary parameters $\dot{q}''_{\text{flux}}$ and T_{AST}, i.e.

$$q''_{\text{flux}} = \varepsilon \cdot \sigma\left(T_{AST}^4 - T_\infty^4\right) + h(T_{\text{AST}} - T_\infty) \tag{4.40}$$

Notice that relation between $\dot{q}''_{\text{flux}}$ and T_∞ is unambiguous and independent on the surface temperature T_s.

For an ambient temperature $T_\infty = 20$ °C and three combinations of ε and h, the relations between AST T_{AST} and the "heat flux" $\dot{q}''_{\text{flux}}$ are shown in Fig. 4.10. As an example for $\varepsilon = 1.0$ and $h = 10$ W/(m^2 K), a "heat flux" $\dot{q}''_{\text{flux}} = 10\,\text{kW/m}^2$ corresponds to an AST $T_{AST} \approx 330$ °C. However, if instead the convection heat transfer coefficient $h = 20$ W/(m^2 K) then the corresponding AST is reduced to $T_{AST} \approx 280$ °C. Thus the assumed values of ε and h have a significant influence on the relation between T_{AST} and $\dot{q}''_{\text{flux}}$.

4.4.4 Calculation of Time Constants for Bodies Exposed to Mixed Boundary Conditions

The concept of adiabatic heat transfer resistance as defined in Eq. 4.37 may be used to calculate the temporal response characteristics (time constants) of bodies exposed to radiation and simultaneously to convection. The time constant for

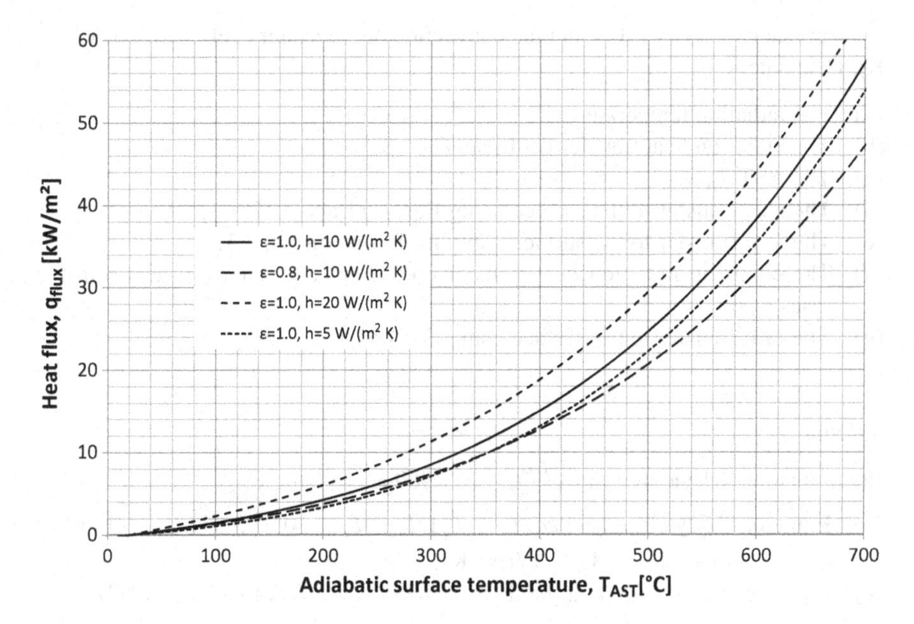

Fig. 4.10 Relation between adiabatic surface temperature T_{AST} and heat flux $\dot{q}''_{flux}$ to a surface at ambient temperature (20 °C) for various combinations of surface emissivities ε and convection heat transfer coefficients *h*

bodies exposed to uniform temperatures is described in, e.g. Sect. 3.1. As a general rule the time constant of bodies exposed to radiation decreases significantly when the temperature level increases as in many FSE scenarios.

Example 4.3 The maximum incident radiation from the sun at the earth's surface perpendicular to the sun's rays is approximately 1 kW/m^2. What is the equilibrium temperature of perfectly insulated surface perpendicular to the rays when

(a) The convection is negligible.
(b) The air temperature is 20 °C, the convection heat transfer coefficient is 10 W/(m^2 K) and the surface equal unity.

Assume a surface emissivity independent of the wavelength.

Solution

(a) Equation 4.1, Table 4.2 or Fig. 4.1 yields the equilibrium temperature 364 K or 91 °C.
(b) Equation 4.22 or Fig. 4.7a yields $T_r = 91$ °C and h/ε = 10 W/(m^2 K) and the equilibrium temperature (AST) 55 °C.

Example 4.4 A surface with a temperature of $T_s = 200$ °C is exposed to an incident radiation of $\dot{q}''_{inc} = 50\,\text{kW/m}^2$ and a gas temperature of $T_g = 150$ °C.

The surface emissivity is 0.9 and the convection heat transfer coefficient 10 W/(m^2 K). Calculate

(a) The radiation temperature.
(b) The radiation heat transfer coefficient.
(c) The AST T_{AST}.
(d) The heat transfer to the surface using Eqs. 4.11 and 4.12, respectively.
(e) The heat transfer to the surface applying T_{AST}using Eq. 4.31.
(f) The adiabatic radiation heat transfer coefficient and adiabatic heat transfer resistance.
(g) The heat transfer to the surface using Eq. 4.34.

Solution

(a) Equation 4.1 yields $T_r = \sqrt[4]{\frac{50 \cdot 10^3}{5.67 \cdot 10^{-8}}} = 969\ \text{K} = 696\ °\text{C}$.
(b) Equation 4.5 yields $h_r = \varepsilon\sigma(T_r^2 + T_s^2)(T_r + T_s) = 0.9 \cdot 5.67 \cdot 10^{-8} \cdot [969^2 + 473^2] \cdot [969 + 473] = 85.5\ \text{W}/(\text{m}^2\ \text{K})$.
(c) Equation 4.21 or Eq. 4.22 yields by iteration $T_{AST} = 940\ \text{K}(= 667\ °\text{C})$.
(d) Equation 4.11 yields $\dot{q}''_{tot} = \varepsilon(\dot{q}''_{inc} - \sigma T_s^4) + h_c(T_g - T_s) = 0.9 \cdot (50 \cdot 10^3 - 5.67 \cdot 10^{-8} \cdot 473^4) + 10 \cdot (150 - 200) = 42{,}440 - 500 = 41{,}940\ \text{W}/\text{m}^2$, or alternatively Eq. 4.12 yields $\dot{q}''_{tot} = \varepsilon\sigma(T_r^4 - T_s^4) + h_c(T_g - T_s) = 0.9 \cdot 5.67 \cdot 10^{-8} \cdot (969^4 - 473^4) + 10 \cdot (150 - 200) = 42{,}440 - 500 = 41{,}900\ \text{W}/\text{m}^2$.
(e) Equation 4.31 yields $\dot{q}''_{tot} = \varepsilon\sigma(T_{AST}^4 - T_s^4) + h_c(T_{AST} - T_s) = 0.9 \cdot 5.67 \cdot 10^{-8} \cdot (940^4 - 473^4) + 10 \cdot (940 - 473) = 41{,}900\ \text{W}/\text{m}^2$.
(f) Equation 4.36 yields $h_r^{AST} = 0.47 \cdot 0.9 \cdot 4 \cdot 5.67 \cdot 10^{-8} \cdot 940^3 = 80\ \text{W}/(\text{m}^2\ \text{K})$. Then $h_{tot}^{AST} = 80 + 10 = 90\ \text{W}/(\text{m}^2\ \text{K})$. $R_{tot}^{AST} = 1/h_{tot}^{AST} = 0.011\ (\text{m}^2\ \text{K})/\text{W}$.
(g) Equation 4.34 yields $\dot{q}''_{tot} = 90 \cdot (940 - 473) = 42{,}000\ \text{W}/\text{m}^2$.

Example 4.5 A wall consists of a wooden panel with a thickness of $d_{wood} = 25$ mm and a conductivity $k_{wood} = 0.1\ \text{W}/(\text{m} \cdot \text{K})$ and a mineral wool insulation with a thickness $d_{ins} = 100$ mm and a conductivity $k_{ins} = 0.02\ \text{W}/(\text{m} \cdot \text{K})$ as shown in Fig. 4.11. The surface temperatures $T_1 = 100\ °\text{C}$ and $T_3 = 20\ °\text{C}$. Calculate the heat flux $\dot{q}''$ through the assembly and the temperature T_2.

Solution

The heat flow through the assembly

$$\dot{q}'' = \frac{T_1 - T_3}{(R_1 + R_2)} = \frac{80}{\left[\left(\frac{0.012}{0.1}\right) + \left(\frac{0.05}{0.02}\right)\right]} = \frac{80}{[0.12 + 2.5]} = 30.5\,\text{W}/\text{m}^2$$

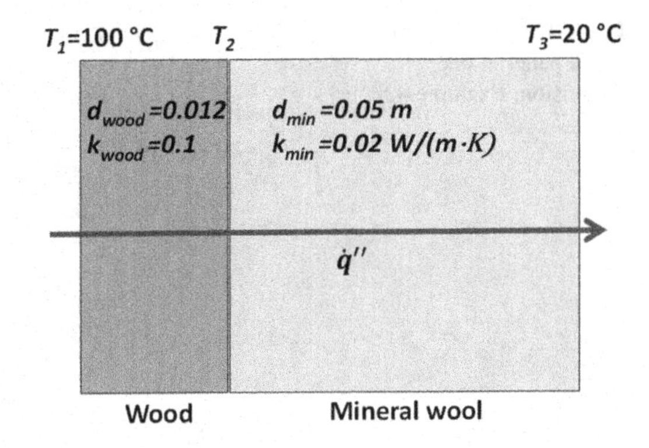

Fig. 4.11 Properties of insulated wooden wall, Example 4.7

$$
\begin{aligned}
T_2 &= T_1 - \frac{R_1}{R_1 + R_2}(T_1 - T_3) = 100 - \frac{0.12}{(0.12 + 2.5)}(100 - 20) \\
&= 100 - 3.66 = 96.3\,°\mathrm{C}
\end{aligned}
$$

Notice that the thermal resistance of the wood panel is much small than that of the insulation and therefore the interface temperature becomes close to that of the wood panel.

Example 4.6 A 30-mm-thick steel sheet is exposed to a gas temperature T_g =500 °C on one side and 20 °C on the other. Calculate the heat flux through the sheet and its surface temperatures. Assume a heat transfer coefficient $h = 100$ W/(m^2 K) on the hot side (1) and 20 W/(m^2 K) on the other (side 2). The steel conductivity $k = 50$ W/(m K).

Solution

The thermal resistance over a unit area is the sum of the heat transfer resistance and conductive resistance. Thus the heat flux through the panel $\dot{q}'' = \frac{T_1 - T_2}{R_{h1} + R_{st} + R_{h2}} = \frac{500 - 20}{\left[\left(\frac{1}{100}\right) + \left(\frac{0.030}{50}\right) + \left(\frac{1}{20}\right)\right]} = \frac{480}{(0.01 + 0.0006 + 0.05)} = 7921\ \mathrm{W/m^2}$ and, e.g. according to Eq. 2.12 $T_1 = \frac{500 \cdot (0.0006{+}0.05) + 20 \cdot 0.01}{(0.01{+}0.0006{+}0.05)} = 465$ °C and $T_2 = \frac{500 \cdot 0.05{+}20 \cdot (0.0006{+}0.01)}{(0.01{+}0.0006{+}0.05)} = 416$ °C. Notice that the temperature change over the steel sheet is relatively small as thermal heat transfer resistances dominates.

Example 4.7 Calculate the surface temperatures T_1 at a steel sheet surface, see Fig. 4.12, when the heat flux through the steel sheet is $\dot{q}'' = 5\ \mathrm{kW/m^2}$. The panel is 25 mm thick and the temperature on the other side $T_2 = 20$ °C. Assume the conductivity of steel $k = 50$ W/(m K).

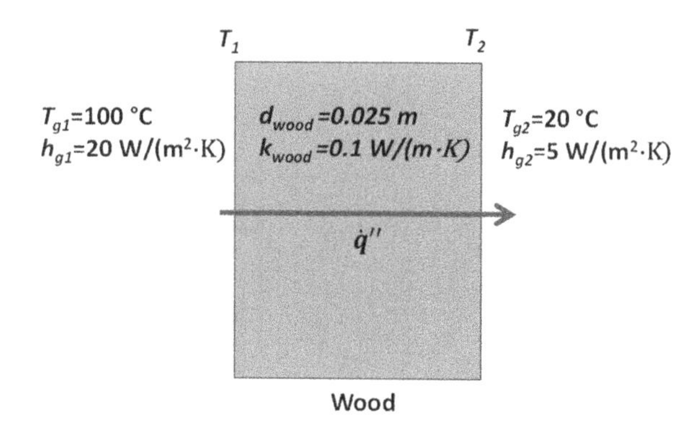

Fig. 4.12 Properties of wood panel in one dimension, Example 4.8

Solution

$T_1 = T_2 + \left(\frac{0.025}{50} \cdot 5000\right) = 20 + 2.5 = 22.5\,°\text{C}.$

Example 4.8 Calculate the surface temperatures T_1 and T_2 of the wood panel when surrounded by temperature $T_{g1} = 100\ °\text{C}$ and $T_{g2} = 20\ °\text{C}$ on the left and the right side, respectively. Consider heat transfer by convection assuming the heat transfer coefficients $h_1 = 20\ \text{W}/(\text{m}^2\ \text{K})$ on the hot side and $h_2 = 5\ \text{W}/(\text{m}^2\ \text{K})$ on the cool side. The wood panel is assumed to be 25 mm thick and have a thermal conductivity $k_{wood} = 0.1$ W/(m K).

Solution Calculate the total thermal resistance over a unit area $R_{tot} = R_h + R_k + R_h = 1/20 + 0.025/0.1 + 1/5 = 0.50\ \text{K/W}$. According to Eq. 2.9 $T_1 = \frac{100\cdot(0.025/0.1+1/5)+20\cdot 1/20)}{0.5} = 92$ and $T_2 = \frac{100\cdot 0.025/0.1+20\cdot(1/20+.025/0.1)}{0.5}$. Alternatively the heat flux $\dot{q}'' = \frac{100-20}{0.5} = 160\text{W}/\text{m}^2$ at the boundaries two heat balance equations can be established: $(100 - T_1)/0.05 = 160$ and $(T_2 - 20)/0.2 = 160$ yielding $T_1 = 92\ °\text{C}$ and $T_2 = 52\ °\text{C}$.

Example 4.9 Calculate the net heat transfer by radiation $\dot{q}''_{rad}$ to a surface at a temperature T_s and an emissivity of 0.9 when exposed to an incident radiation heat flux of 20 kW/m². (a) Assume $T_s = 20\ °\text{C}$, (b) assume $T_s = 500\ °\text{C}$ and (c) what is the exposure black body radiation temperature T_r?

Solution

According to Eq. 1.15:

(a) $\dot{q}''_{rad} = 0.9\left[20{,}000 - \sigma(20+273)^4\right] = 17{,}610 \approx 17.6\ \text{kW}/\text{m}^2$

(b) $\dot{q}''_{rad} = 0.9\left[20{,}000 - \sigma(500+273)^4\right] = -220 \approx -0.220\text{kW}/\text{m}^2$

(c) According to Eq. 4.1 $T_r = \left[\frac{\dot{q}''_{inc}}{\sigma}\right]^{1/4} = 771\ \text{K} = 498\ °\text{C}$

Chapter 5
Heat Transfer by Radiation

Heat transfer by thermal radiation is transfer of heat by electromagnetic waves. It is different from conduction and convection as it requires no matter or medium to be present. The radiative energy will pass perfectly through vacuum as well as clear air. While the conduction and convection depend on temperature differences to approximately the first power, the heat transfer by radiation depends on the differences of the individual body surface temperatures to the fourth power. Therefore the radiation mode of heat transfer dominates over convection at high temperature levels as in fires. Numerical applications of radiation heat transfer in FSE are outlined in Sect. 4.1.

The description below is mainly taken from [11]. The surfaces are generally assumed to be grey, which means they absorb and emit radiation that is a fraction of black body radiation in all directions and over all wavelengths. Hence the hemispherical absorptivity/emissivity of a surface is assumed to be independent of the nature of the incident radiation and of the spectral properties of, e.g. a fire.

The upper limit of the heat flux leaving a *black body* surface by radiation is according to the *Stefan–Boltzmann law*

$$\dot{q}''_{bb,emi} = \sigma \cdot T_s^4 \tag{5.1}$$

where σ is the *Stefan–Boltzmann constant* ($\sigma = 5.670 \cdot 10^{-8}$ [W/(m^2 K)]) and T_s is the absolute surface temperature [K]. Figure 4.1 can be used to calculate the emitted heat by radiation from a black surface vs. temperature in Kelvin, K, according to Eq. 5.1 or vs. temperature in degree Celsius, °C.

The heat flux $\dot{q}''_{emi}$ leaving a real surface is, however, less than that of a black body at the same temperature:

$$\dot{q}''_{emi} = \varepsilon_s \cdot \sigma \cdot T_s^4 \tag{5.2}$$

where ε_s is the *emissivity* of the surface.

U. Wickström, *Temperature Calculation in Fire Safety Engineering*,
DOI 10.1007/978-3-319-30172-3_5

The *incident radiation* $\dot{q}''_{inc}$ to a surface may originate from various sources. When it includes radiation irrespective of sources it is sometimes called *irradiance*.

Only a fraction of the incident radiation $\dot{q}''_{abs}$ will be absorbed by a surface, i.e.

$$\dot{q}''_{abs} = \alpha_s \cdot \dot{q}''_{inc} \tag{5.3}$$

where α is the *absorptivity* of the surface. The rest of the incident radiation is reflected $\dot{q}''_{ref}$ or transmitted through the surface. The latter term is small for most materials and is neglected in the presentation below. Hence the reflected radiation heat flux becomes

$$\dot{q}''_{ref} = (1 - \alpha_s) \cdot \dot{q}''_{inc} \tag{5.4}$$

The net rate of heat flux to a surface by radiation then becomes:

$$\dot{q}''_{rad} = \dot{q}''_{abs} - \dot{q}''_{emi} \tag{5.5}$$

or after inserting Eqs. 5.2 and 5.3 and given the *Kirchhoff's identity* $\alpha_s = \varepsilon_s$, the heat flux to a surface by radiation becomes (Fig. 5.1)

$$\dot{q}''_{rad} = \varepsilon_s \left(\dot{q}''_{inc} - \sigma T_s^4 \right) \tag{5.6}$$

The incident radiation or the irradiation on a surface is emitted by other surfaces and/or by surrounding masses of gas and in case of fire by flames and smoke layers. The emissivity and absorptivity of gas masses and flames increase with depth and becomes therefore more important in large scale fires than in, e.g. small scale experiments, see Sect. 5.3. In real fires surfaces are exposed to radiation from a large number of sources, surfaces, flames, gas masses, etc., of different temperatures and emissivities and the incident radiation is in general very complicated to model. If absorption from any gases is neglected, and if the target surface is small and therefore the contributions of reflections and re-radiation are neglected, the incident radiation to the surface can be approximated as the sum of the contributions $\dot{q}''_{inc,i}$ from a number of external sources:

$$\dot{q}''_{inc} = \sum_i \dot{q}''_{inc,i} \tag{5.7}$$

When the source number i is a surface with a uniform temperature T_i the contribution is

$$\dot{q}''_{inc,i} = \varepsilon_i \cdot F_i \cdot \sigma \cdot T_i^4 \tag{5.8}$$

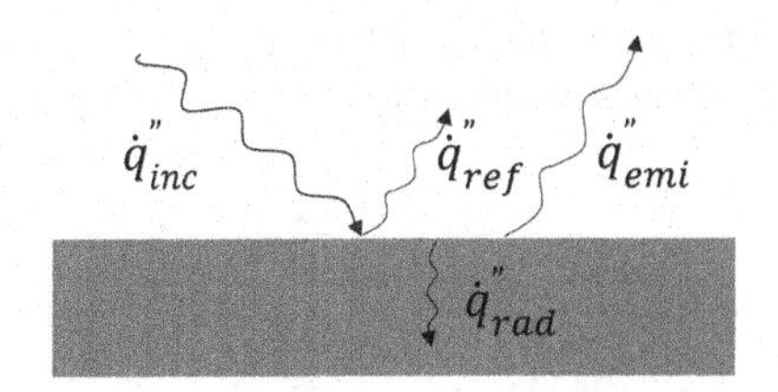

Fig. 5.1 The heat transfer by radiation to a surface depends on incident radiation and the absolute temperature and emissivity of the surface

where ε_i is the emissivity of the *i*th source. F_i is the corresponding view factor as defined in more detail in Sect. 5.2. Like the emissivity it always has values between 0 and 1.

The incident radiation may also be written as a function of the *black body incident radiation temperature*, the *black body radiation temperature* or just the *radiation temperature* defined by the identity $\dot{q}''_{inc} \equiv \sigma \cdot T_r^4$ (Eq. 1.17).

Thus T_r is a weighted average of the surrounding surface temperatures which can be obtained by combining Eqs. 5.7 and 5.8 as

$$T_r \equiv \sqrt[4]{\left[\left(\sum_i \varepsilon_i \cdot F_i \cdot T_i^4\right)/\sigma\right]} \tag{5.9}$$

T_r can also be defined as the temperature a surface will get which is in radiation equilibrium with the incident flux, i.e. no heat is transferred neither by convection nor by conduction from that surface. Compare with the concept of *adiabatic surface temperature*, as described in Sect. 4.4, which is the surface temperature when the *net* radiation $\dot{q}''_{rad}$ is in equilibrium with the convection heat flux $\dot{q}''_{con}$.

The net radiation heat flux is obtained by subtracting the emitted radiation according to Eq. 5.2 from the absorbed:

$$\dot{q}''_{rad} = \varepsilon_s \left[\sum_i \left(\dot{q}''_{inc,i}\right) - \sigma \cdot T_s^4\right] \tag{5.10}$$

where ε_s is the emissivity/absorptivity of the target surface.

The surfaces emissivities of some materials are given in Table 5.1. In general the emissivity of all real/technical materials is in the range of 0.75–0.95 except shiny steel where the emissivity can be considerably lower. It depends on the temperature of heat source and decreases in general with the heat source grey body temperature. Typically values of the absorptivity of plywood drop from 0.86 to 0.76 when the source temperature increases from 674 to 1300 K [12]. The corresponding value for radiation emitted from the sun (5777 K) is as low as 0.40. Eurocode 2 [6] and Eurocode 3 [3] recommend 0.7 for concrete and steel, respectively. The choice of emissivity is primarily of importance when calculating temperature of fire-exposed bare steel structures. For lightweight insulating materials the surface temperature adapts quickly to the exposure conditions and therefore the heat transfer conditions, expressed by the heat transfer coefficient, are negligible for the temperature development.

Table 5.1 Surface emissivity of some common materials

Material	Emissivity, ε
Concrete	0.8[a]
Steel	0.7[a]
Min.wool	0.9
Paint	0.9
Red bricks	0.9
Wood	0.9
Sand	0.9
Rocks	0.9
Water	0.96

The values are uncertain and should be taken as indicative
[a]From Eurocode

5.1 Radiation Between Two Parallel Planes and Radiation Shields

When two infinite parallel plates as shown in Fig. 5.2a are considered, the radiation view factor is unity as all the heat emitted or reflected at one surface will incident on the other. Some of that heat will be absorbed and some will be reflected back to the opposite surface. The net heat flux from surface one to two may be calculated as

$$\dot{q}''_{rad,1-2} = \varepsilon_{res}\sigma\left(T_1^4 - T_2^4\right) \tag{5.11}$$

where the *resultant emissivity* is defined as

$$\varepsilon_{res} = \frac{1}{\frac{1}{\varepsilon_1} + \frac{1}{\varepsilon_2} - 1} \tag{5.12}$$

Radiation exposure can be considerably reduced by a radiation shield. Figure 5.2b shows an example where a shield is mounted between two surfaces. The shield has no thermal resistance, i.e. its both sides have the same temperature. The radiation heat flux rate between surface 1 to the shield must equal the flux rate between the shield and the surface 2, i.e. $\dot{q}''_{rad} = \dot{q}''_{1-sh} = \dot{q}''_{sh-2}$ or $\dot{q}''_{rad} = \varepsilon_{res,1-sh}\sigma\left(T_1^4 - T_{sh}^4\right) = \varepsilon_{res,sh-2}\sigma\left(T_{sh}^4 - T_2^4\right)$. Thus the shield temperature to the fourth power can be derived as

$$\mathrm{T}_{\mathrm{sh}}^4 = \frac{\varepsilon_{\mathrm{res,1-sh}}\mathrm{T}_1^4 + \varepsilon_{\mathrm{res,sh-2}}\mathrm{T}_2^4}{\left(\varepsilon_{\mathrm{res,1-sh}} + \varepsilon_{\mathrm{res,sh-2}}\right)}. \tag{5.13}$$

and if all the four surface emissivities defined in Fig. 5.2b are equal to ε_s then

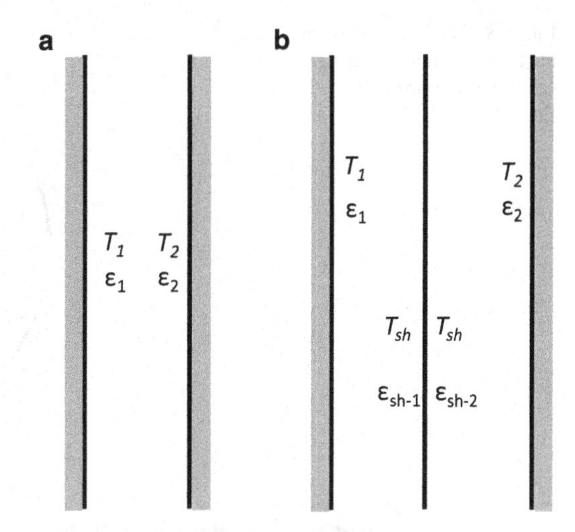

Fig. 5.2 Radiation heat transfer between two large parallel plates, without (**a**) and with (**b**) a radiation shield

$$T_{sh}^4 = \frac{T_1^4 + T_2^4}{2} \tag{5.14}$$

and radiation heat flux between the surfaces 1 and 2 becomes

$$\dot{q}''_{rad} = \frac{\varepsilon_s}{(2 - \varepsilon_s)} \sigma\left(T_1^4 - T_2^4\right)/2 \tag{5.15}$$

Equation 5.14 implies that in the case of equal emissivities, the temperature of, e.g. a fire radiation shield is closer to the higher (fire) temperature than to the lower (ambient) temperature. Under the same conditions Eq. 5.15 shows the heat flux by radiation is reduced by 50 %. A reduction of the common emissivity will reduce heat transfer correspondingly although it will not change the temperature of the shield as according to Eq. 5.14 the temperature of the shield is independent of the emissivity.

Equation 5.15 may be extended to problems involving multiple radiation shields with all surface emissivities being equal to ε_s. Then with N shields the heat flux $\dot{q}''_{rad,N}$ becomes

$$\dot{q}''_{rad,N} = \frac{1}{(N+1)} \dot{q}''_{rad,0} = \frac{1}{(N+1)} \cdot \frac{1}{\frac{2}{\varepsilon_s} - 1} \sigma\left(T_1^4 - T_2^4\right) \tag{5.16}$$

where $\dot{q}''_{rad,0}$ is the radiation heat flux with no shields ($N = 0$) according to Eqs. 5.11 and 5.12 with equal emissivities.

The corresponding formula for the flux between infinitely long concentric cylinders as indicated in Fig. 5.3 is

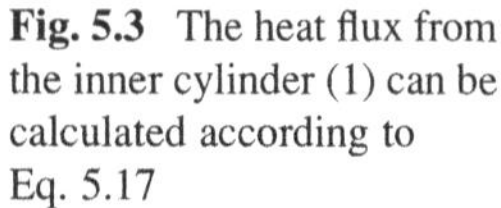

Fig. 5.3 The heat flux from the inner cylinder (1) can be calculated according to Eq. 5.17

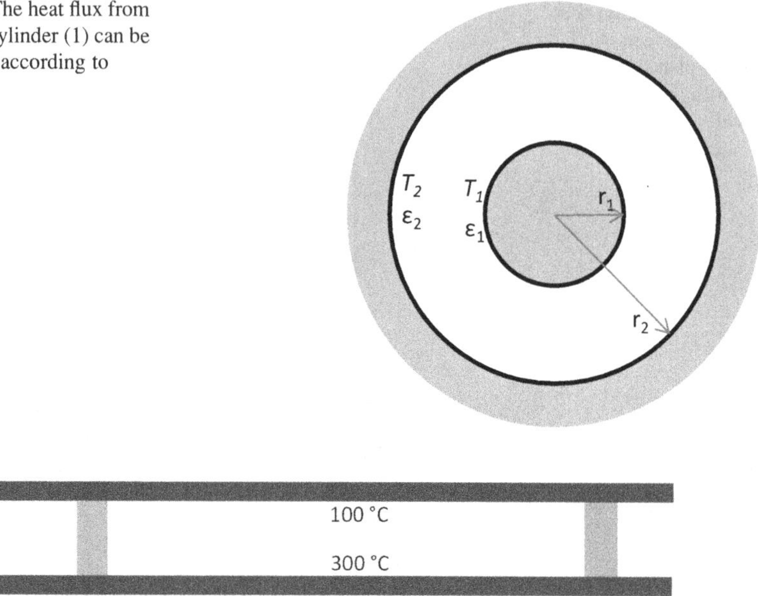

Fig. 5.4 Uninsulated wooden stud wall

$$\dot{q}''_{rad,1-2} = \frac{\sigma(T_1^4 - T_2^4)}{\frac{1}{\varepsilon_1} + \frac{r_1}{r_2} \cdot \left(\frac{1}{\varepsilon_2} - 1\right)} \tag{5.17}$$

where r_1 and r_2 are the inner radii and ε_1 and ε_2 the corresponding surface emissivities. Notice that if $r_1 \ll r_2$ or more generally for a small object in a large cavity the heat flux from the inner object becomes independent of the outer surface emissivity, i.e.

$$\dot{q}''_{rad,1-2} = \varepsilon_1 \cdot \sigma(T_1^4 - T_2^4) \tag{5.18}$$

Example 5.1 The inside surfaces of the boards of an uninsulated wooden stud wall as shown in Fig. 5.4 have the temperatures 300 and 100 °C, respectively. Calculate the heat flux by radiation and convection between the board surfaces. The distance between the boards is 100 mm and between the studs 600 mm. The emissivity of the board surfaces is 0.9. Assume one-dimensional heat flux.

Solution Equations 5.11 and 5.12 yield $\dot{q}''_{rad} = \frac{1}{\frac{1}{0.9}+\frac{1}{0.9}-1} \cdot 5.67 \cdot 10^{-8} \cdot \left[(300+273)^4 - (100+273)^4\right] = 4104\ \mathrm{W/m^2}$. For the convection heat transfer, see Example 6.7. Thus $\dot{q}''_c = 644\ \mathrm{W/m^2}$ and the total heat flux $\dot{q}''_{tot} = (4104 + 644)\ \mathrm{W/m^2} = 4748\ \mathrm{W/m^2}$. Notice in this case heat flux by convection is less than 5 % of the total.

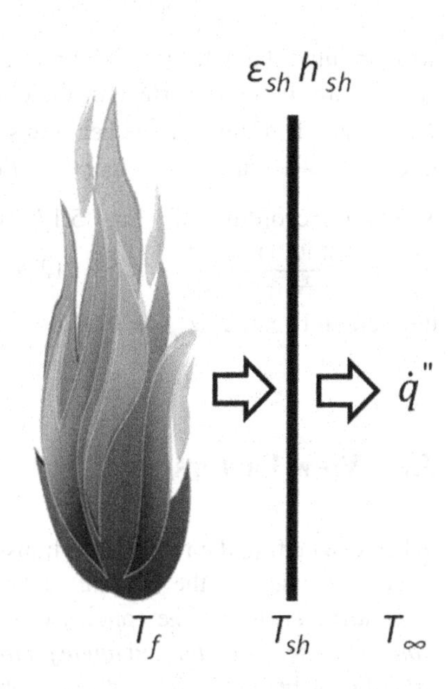

Fig. 5.5 Radiation shield

Example 5.2 The radiation from a large flame at a temperature $T_f = 1000$ K is reduced by a metal radiation shield having no thermal conduction resistance (Fig. 5.5). The ambient gas temperature $T_\infty = 300$ K. Calculate the temperature T_{sh} of the shield and the relative reduction of the radiation level $\dot{q}''$ by the shield. Assume

(a) $\varepsilon_s = 1$ and $h = 0$.
(b) $\varepsilon_s = 0.5$ and $h = 0$.
(c) $\varepsilon_s = 1$ and $h = 6$ W/m^2 K.

Solution

(a) All "surfaces" have an emissivity equal to unity. Thus according to Eq. 5.14 $T_{sh} = \left(\sqrt[4]{1000^4 + 293^4}\right)/2 = 842$ K and $\dot{q}'' = 5.67 \cdot 10^{-8} \cdot 842^4 = 28.5$ W/m^2, i.e. a reduction by 50 %.
(b) $T_{sh} = \left(\sqrt[4]{1000^4 + 293^4}\right)/2 = 842$ K and $\dot{q}'' = 0.5 \cdot 28.5 = 14.3$ W/m^2, i.e. a reduction by 25 %.
(c) An iteration formula can be derived $T_{sh}^{i+1} = \left[\left(\sqrt[4]{1000^4 + 293^4}\right) - 6/5.67 \cdot 10^{-8} \cdot (T_{sh}^{i} - 293)\right]/2$ yields $T_{sh} = 818$ K and $\dot{q}'' = 5.67 \cdot 10^{-8} \cdot 818^4 = 25.4$ W/m^2, i.e. a reduction by 45 %.

Example 5.3 Gas flows through a long tube of $r_1 = 40$ mm diameter with an outer surface emissivity $\varepsilon_1 = 0.3$. The tube is concentric with an outer insulation tube with an inner diameter $r_2 = 100$ mm and an inner surface emissivity $\varepsilon_2 = 0.8$. In case of fire the inner surface of the outer tube is expected to reach a temperature $T_2 = 1200$ K. Calculate the heat transfer by radiation per metre length $\dot{q}'_{rad}$ to the inner tube when it has a temperature $T_1 = 500$ K.

Solution According to Eq. 5.17 the heat flux to the inner surface is $\dot{q}''_{rad} = \frac{5.67 \cdot 10^{-8}\left(1200^4 - 500^4\right)}{\frac{1}{0.3} + \frac{40}{100} \cdot \left(\frac{1}{0.8} - 1\right)} = 33,212\ \mathrm{W/m^2}$ and the heat transfer by radiation per unit length becomes $\dot{q}'_{rad} = 0.040 \cdot \pi \cdot 33,212 = 4173\ \mathrm{W/m}$.

5.2 View Factors

When calculating the rate of heat transfer by radiation between surfaces, a method is needed whereby the amount of heat being radiated in any direction can be calculated. Therefore the concept *view factor* is introduced. The terms *configuration factor*, *shape factor* and *angle factor* are also used. The physical meaning of the view factor between two surfaces is the fraction of radiation leaving one surface that arrives at the other directly. The symbol $F_{A_1-A_2}$ is used to denote the view factor from a surface A_1 to a surface A_2. The symbol $F_{dA_1-A_2}$ denotes the view factor from an incremental surface dA_1 to a finite surface A_2. View factors defined in this way are functions of size, geometry, position and orientation of the two surfaces. View factors are between zero and unity, and the sum of the view factors of a surface is one.

Thus by definition the radiation leaving a surface A_1 arriving at a surface A_2 is

$$\dot{q}_{inc,1-2} = F_{1-2} A_1 \dot{q}''_{emi,1} \tag{5.19}$$

and similarly the radiation leaving a surface A_2 arriving at a surface A_1 is

$$\dot{q}_{inc,2-1} = F_{2-1} A_2 \dot{q}''_{emi,2} \tag{5.20}$$

A *reciprocity relation* can be derived which reads

$$A_1 F_{1-2} = A_2 F_{2-1} \tag{5.21}$$

and

$$F_{2-1} = \frac{A_1}{A_2} F_{1-2} \tag{5.22}$$

In a more general way for any two surfaces i and j

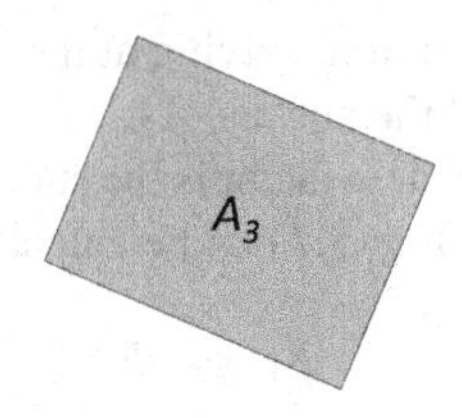

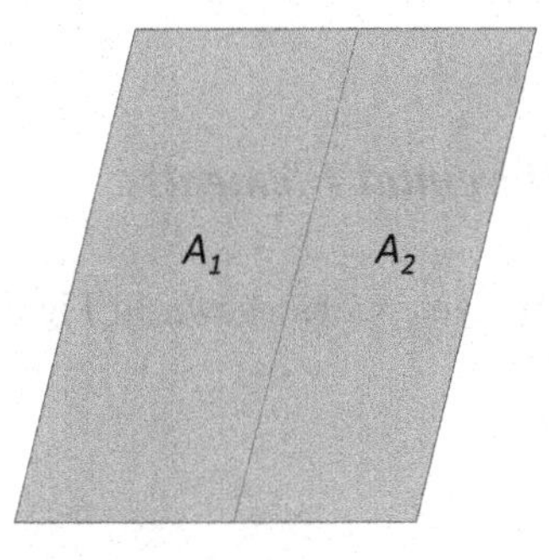

Fig. 5.6 Relations between view factors

$$A_iF_{i-j} = A_jF_{j-i} \tag{5.23}$$

Another useful relation between view factors may be obtained by considering the system shown in Fig. 5.6. The view factor from the surface A_3 to the combined surface $A_{1,2}$ is then

$$F_{3-1,2} = F_{3-1} + F_{3-2} \tag{5.24}$$

That is the total view factor is the sum of its parts. Eq. 5.24 could also be written as

$$A_3F_{3-1,2} = A_3F_{3-1} + A_3F_{3-2} \tag{5.25}$$

and then the reciprocity relations below can be applied:

$$\begin{aligned} A_3F_{3-1,2} &= A_{1,2}F_{1,2-3} \\ A_3F_{3-1} &= A_1F_{1-3} \\ A_3F_{3-2} &= A_2F_{2-3} \end{aligned} \tag{5.26}$$

Equation 5.25 can now be written as

$$A_{1,2}F_{1,2-3} = A_1F_{1-3} + A_2F_{2-3} \tag{5.27}$$

That means the total radiation arriving at the surface A_3 is the sum of the radiations from the surface A_1 and the surface A_2.

The fact that the total view factor is the sum of its parts implies that the view factor F_{1-3} for the surfaces in Fig. 5.7 can be calculated from tabulated view factors as

$$F_{1-3} = F_{1-2,3} - F_{1-2} \tag{5.28}$$

Below some elementary examples are given. A lot more information can be found in textbooks such as [1, 2, 11].

5.2.1 View Factors Between Differential Elements

The view factor between two differential elements as shown in Fig. 5.8 can be obtained as

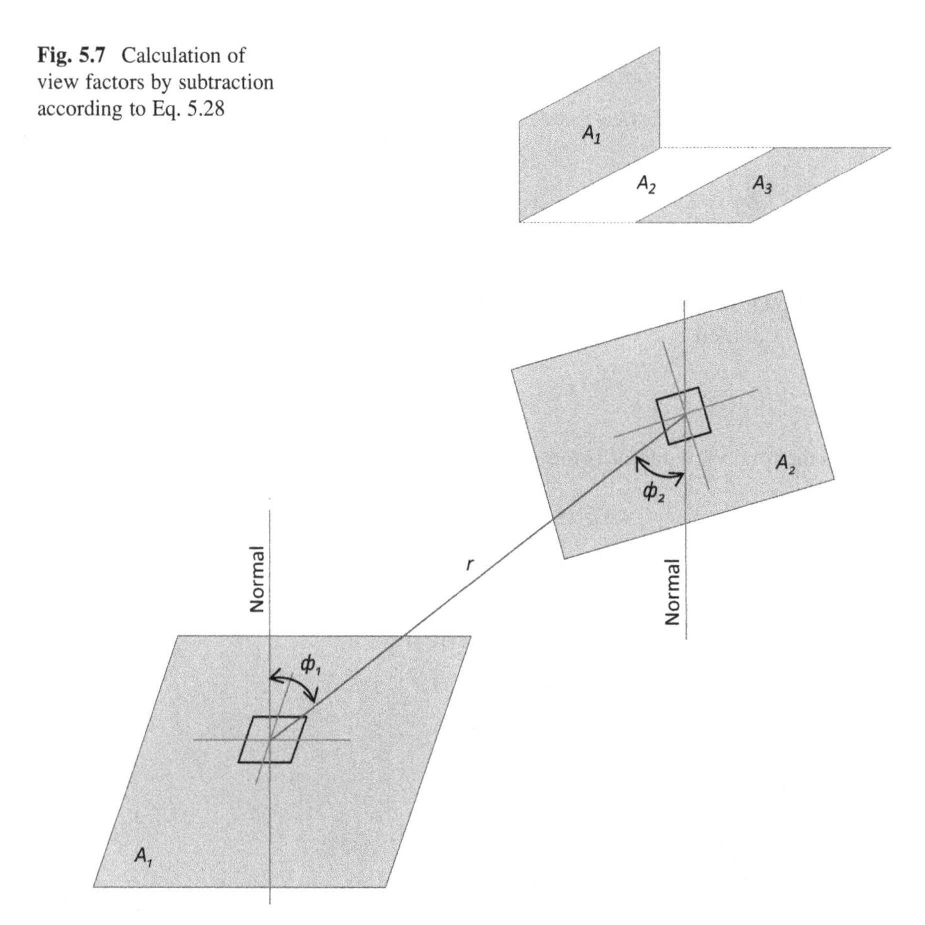

Fig. 5.7 Calculation of view factors by subtraction according to Eq. 5.28

Fig. 5.8 Differential area elements used in deriving view factors according to Eq. 5.29

$$dF_{d1-d2} = \frac{\cos\varphi_1 \ \cos\varphi_2}{\pi r^2} dA_2 \tag{5.29}$$

and correspondingly

$$dF_{d2-d1} = \frac{\cos\varphi_1 \ \cos\varphi_2}{\pi r^2} dA_1 \tag{5.30}$$

The reciprocity relation as given by Eq. 5.21 can be used to derive the equation

$$dF_{d1-d2} dA_1 = dF_{d2-d1} dA_2 = \frac{\cos\varphi_1 \ \cos\varphi_2}{\pi r^2} dA_1 dA_2 \tag{5.31}$$

Now the energy exchange between two black differential elements can be written as

$$d^2\dot{q}_{d1-d2} = \sigma\left(T_1^4 - T_2^4\right) dF_{d1-d2} dA_1 = \sigma\left(T_1^4 - T_2^4\right) dF_{d2-d1} dA_2 \tag{5.32}$$

Then by inserting Eq. 5.29 or Eq. 5.30 the heat exchange between two differential elements becomes

$$d^2\dot{q}_{d1-d2} = \sigma\left(T_1^4 - T_2^4\right) \frac{\cos\varphi_1 \ \cos\varphi_2}{\pi r^2} dA_2 dA_1 \tag{5.33}$$

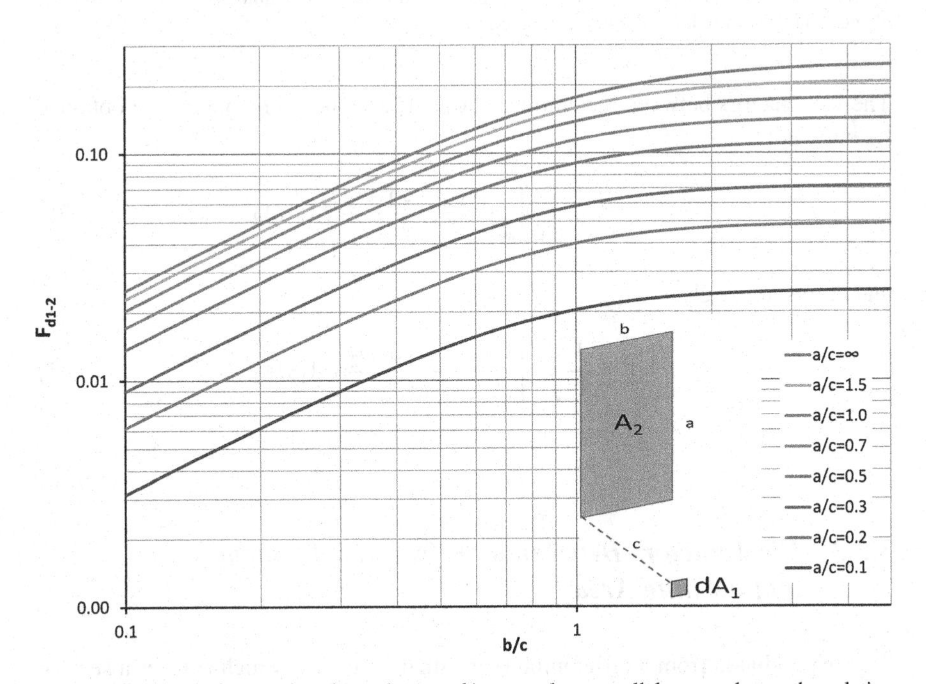

Fig. 5.9 The view factor of a plane element dA_1 to a plane parallel rectangle vs. the relative distances X = a/c and Y = b/c as defined in row 1 of Table 5.2. The normal to the element passes through the corner of the rectangle

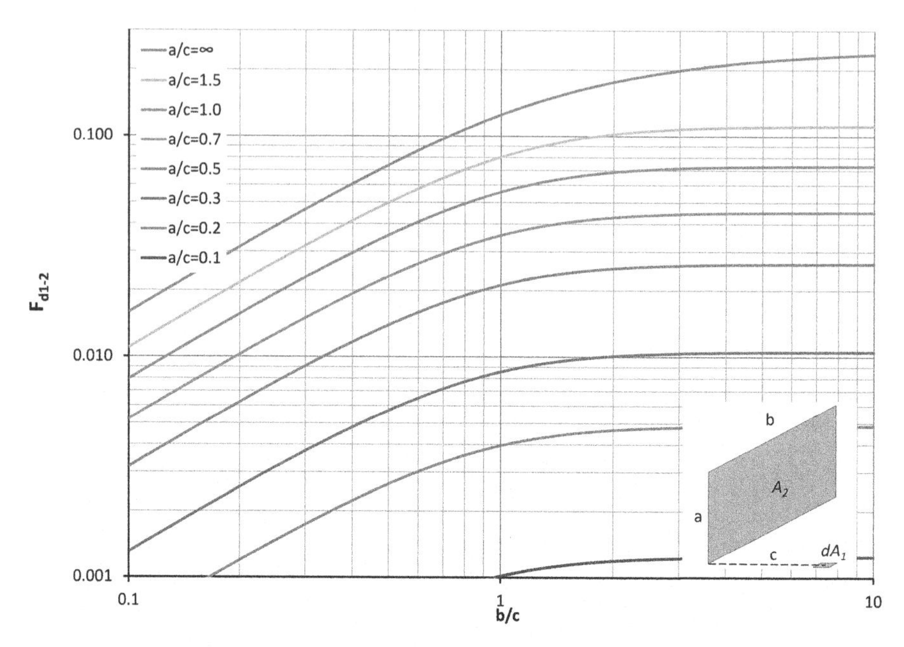

Fig. 5.10 The view factor of a plane element dA_1 to a plane rectangle perpendicular to the element vs. the relative distances X = a/c and Y = b/c as defined in row 2 of Table 5.2. The normal to the element passes through the corner of the rectangle

The view factors between the entire surfaces A_1 and A_2 of Fig. 5.8 can be obtained by integration as

$$F_{1-2} = \frac{1}{A_1} \int_{A_1} \int_{A_2} \frac{\cos \varphi_1 \, \cos \varphi_2}{\pi r^2} dA_1 \, dA_2 \tag{5.34}$$

and correspondingly

$$F_{2-1} = \frac{1}{A_2} \int_{A_1} \int_{A_2} \frac{\cos \varphi_1 \, \cos \varphi_2}{\pi r^2} dA_1 \, dA_2 \tag{5.35}$$

5.2.2 View Factors Between a Differential Element and a Finite Area

The heat radiated from a differential (very small) area dA_1 which reaches a surface A_2 is by the definition of the view factor

Table 5.2 Examples of formulas for calculating the view factors between differential elements and finite areas

1	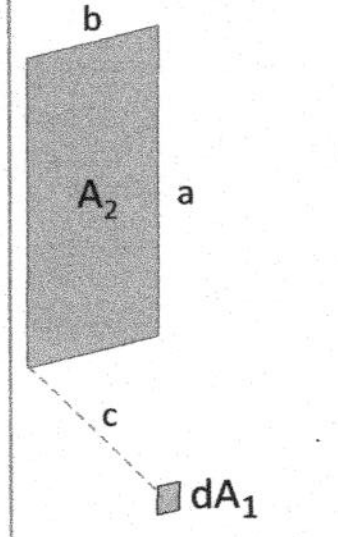	Plane element dA_1 to plane parallel rectangle. Normal to element passes through corner of rectangle, see also Fig. 5.9 $X = \frac{a}{c}$ $Y = \frac{b}{c}$ $F_{d1-2} = \frac{1}{2\pi}\left(\frac{X}{\sqrt{1+X^2}}\tan^{-1}\frac{Y}{\sqrt{1+X^2}} + \frac{Y}{\sqrt{1+Y^2}}\tan^{-1}\frac{X}{\sqrt{1+Y^2}}\right)$
2	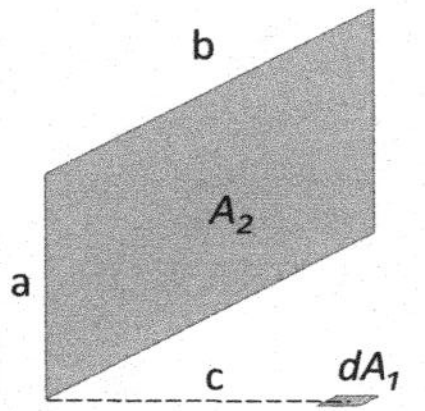	Plane element dA_1 to plane rectangle in plane perpendicular to element. Normal at rectangle corner passes through element, see also Fig. 5.10 $F_{d1-2} = \frac{1}{2\pi}\left[\tan^{-1}\left(\frac{b}{c}\right) - \frac{c}{\sqrt{a^2+c^2}}\tan^{-1}\left(\frac{b}{\sqrt{a^2+c^2}}\right)\right]$
3	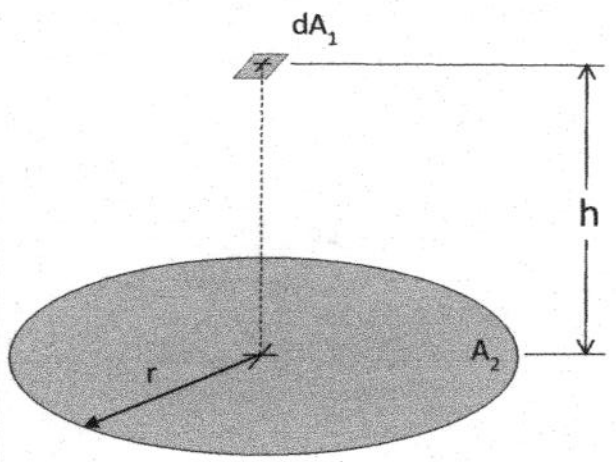	Plane element dA_1 to circular disk in plane parallel to element through centre of disk, see also Fig. 5.12 $F_{d1-2} = \frac{r^2}{h^2+r^2} = \frac{1}{\left(\frac{h}{r}\right)^2+1}$

(continued)

Table 5.2 (continued)

4	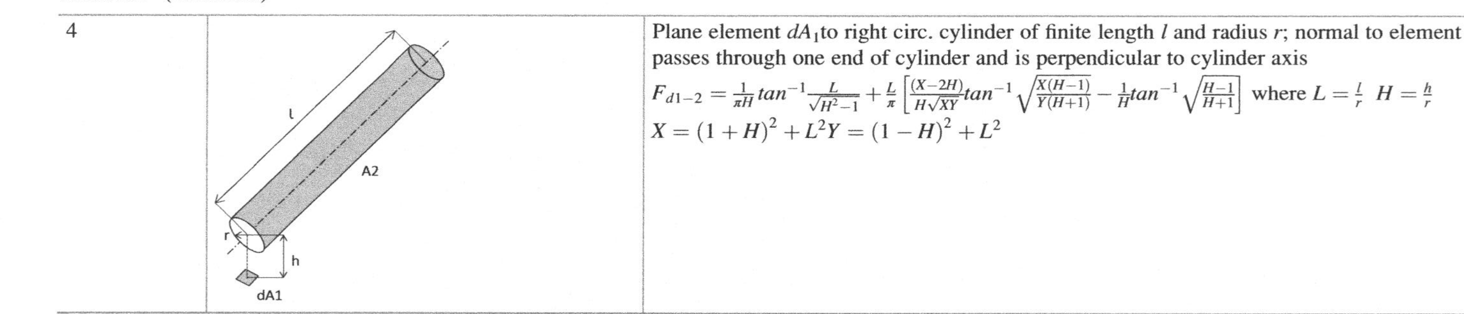	Plane element dA_1 to right circ. cylinder of finite length l and radius r; normal to element passes through one end of cylinder and is perpendicular to cylinder axis $F_{d1-2} = \frac{1}{\pi H} tan^{-1} \frac{L}{\sqrt{H^2-1}} + \frac{L}{\pi} \left[\frac{(X-2H)}{H\sqrt{XY}} tan^{-1} \sqrt{\frac{X(H-1)}{Y(H+1)}} - \frac{1}{H} tan^{-1} \sqrt{\frac{H-1}{H+1}} \right]$ where $L = \frac{l}{r}$ $H = \frac{h}{r}$ $X = (1+H)^2 + L^2 Y = (1-H)^2 + L^2$

$$\dot{q}_{inc,d1-2} = dA_1 F_{d1-2} \dot{q}''_{emi,1} = dA_1 F_{d1-2} \cdot \varepsilon_2 \cdot \sigma \cdot T_1^4 \tag{5.36}$$

where ε_2 is the emissivity of the emitting surface. (Any reflected radiation is here neglected). Examples of such configurations can be seen in Table 5.2. Similarly the heat radiated by A_2 and reaching dA_1 is

$$\dot{q}_{inc,2-d1} = A_2 F_{2-d1}\, q''_{emi,2} = A_2 F_{2-d1} \cdot \varepsilon_2 \cdot \sigma \cdot T_2^4 \tag{5.37}$$

The reciprocity relation according to Eq. 5.21 then yields

$$\dot{q}_{inc,2-d1} = dA_1 F_{d1-2}\, q''_{emi,2} \tag{5.38}$$

and the incident radiation flux to the differential area surface becomes

$$\dot{q}''_{inc,2-d1} = F_{d1-2} \cdot q''_{emi,2} = F_{d1-2} \cdot \varepsilon_2 \cdot \sigma \cdot T_2^4 \tag{5.39}$$

This is the most commonly applied formula version in FSE as it can be used to estimate the incident radiant flux at point where it is expected to be most severe. When several finite surfaces from 2 to n are radiating on an infinite area dA_1, the total incident radiation can be written as

$$\dot{q}''_{inc,(2-n)-d1} = \sigma \sum\nolimits_{i=2}^{n} F_{d1-i} \cdot \varepsilon_i \cdot T_i^4 \tag{5.40}$$

In principle when calculating the total incident radiation to a surface the incident radiation from all angles must be included. Observe that the sum of the view factors is unity. Usually, however, only the contributions from the hot areas such as flame surfaces need be considered as the contributions from, for example, surface at ambient temperature are negligible.

When several surfaces are involved the view factor may be obtained by adding up the contributions from the individual surfaces according to Eq. 5.27. In the case shown in Fig. 5.11 the view factor $F_{dA_1-A_{2-5}}$ between the differential area dA_1 and the entire finite area A_{2-5} may be calculated as

$$F_{d1-(2-5)} = F_{d1-2} + F_{d1-3} + F_{d1-4} + F_{d1-5} \tag{5.41}$$

View factors of various configurations can be found in textbooks such as [1, 2] and particularly in [11].

Table 5.2 shows how to calculate view factors for some elementary cases useful in FSE.

Corresponding diagrams of the view factors defined in rows 1–3 of Table 5.2 are shown Fig. 5.12.

Example 5.4 An un-insulated steel door leaf becomes uniformly heated to a temperature of 500 °C during a fire. Calculate the maximum incident radiation

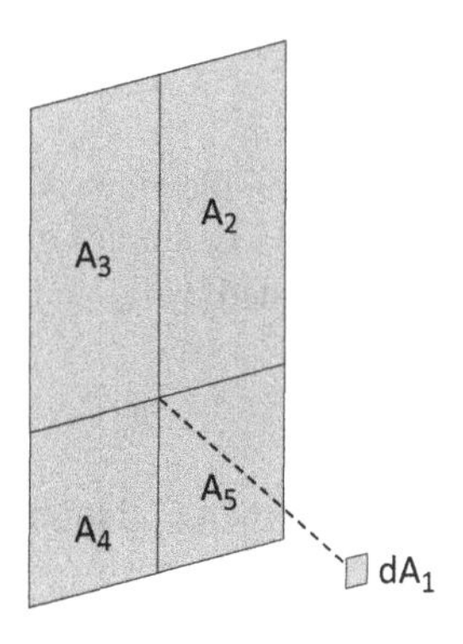

Fig. 5.11 The view factor can be obtained by summing up the contributions of several areas as given by Eq. 5.41

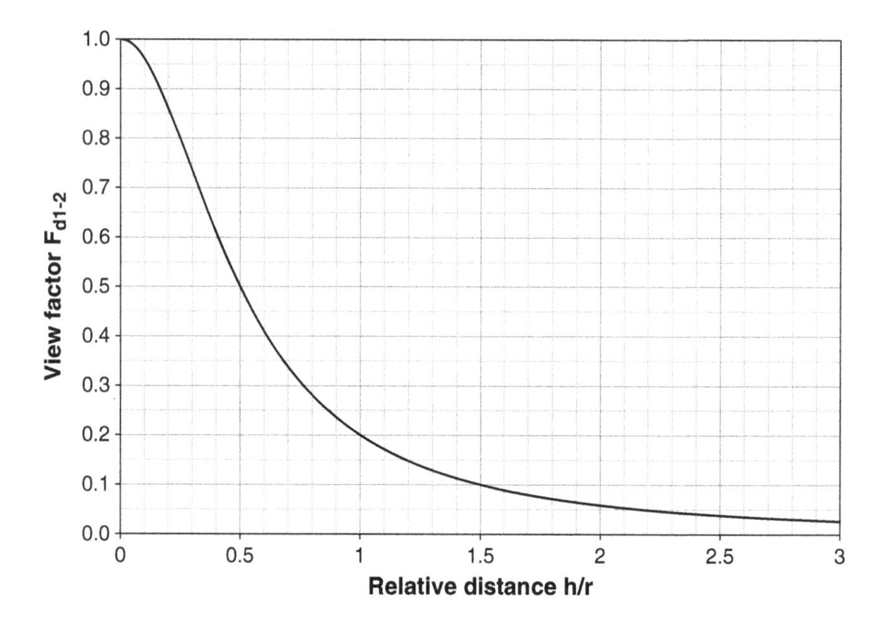

Fig. 5.12 The view factor of plane element dA_1 to a circular disk in a plane parallel to the element vs. the relative distance h/r through the centre of the disk as defined in row 3 of Table 5.2

$\dot{q}''_{inc,max}$ to a parallel surface 1 m from the door leaf with dimensions 0.9 m by 2.1 m and an emissivity of 0.9.

Solution The highest incident radiation will be perpendicular to the centre of the door leaf. Then Eq. 5.39 applies. When calculating the view factor the door is divided into four equal areas and the view factor is obtained as the sum of the four contributions. Then according to the first case in Table 5.2, $X = \frac{2.1/2}{1} = 1.05$ and $Y = \frac{0.9/2}{1} = 0.45$ and the total view factor becomes $F_{dA_1-A_2} =$

$4 \cdot \frac{1}{2\pi}\left(\frac{1.05}{\sqrt{1+1.05^2}} tan^{-1}\frac{0.45}{\sqrt{1+1.05^2}} + \frac{0.45}{\sqrt{1+0.45^2}} tan^{-1}\frac{1.05}{\sqrt{1+0.45^2}}\right) = 4 \cdot 0.085 = 0.34$. The view factor for one-quarter of the door leaf (0.085) can alternatively be obtained from Fig. 5.9. The maximum incident radiation $\dot{q}''_{inc,max} = 0.34 \cdot 0.9 \cdot 5.67 \cdot 10^{-8} \cdot 773^4 = 6200\ \mathrm{W/m^2}$ (corresponding to a black body radiation temperature of 574 K = 301 °C).

5.2.3 View Factors Between Two Finite Areas

In analogy with view factors between a differential element and a finite area (Eq. 5.36) may the heat flow (with units [W]) from one finite area to another be calculated as

$$\dot{q}_{inc,1-2} = A_1 F_{1-2}\, q''_{emi,1} = A_1 F_{1-2}\, \varepsilon_1 \cdot \sigma \cdot T_1^4 \tag{5.42}$$

The net exchange from A_1 to A_2 assuming black isothermal surfaces ($\varepsilon = 1$) is

$$\dot{q}_{2\leftrightarrow 1} = A_1 F_{1-2}\, \sigma\left(T_1^4 - T_2^4\right) \tag{5.43}$$

Table 5.3 shows two examples on how to calculate view factors between finite surfaces, two parallel circular disks with centres along the same normal and two infinitely long plates of unequal widths having a common edge at an angle of 90 °C to each other.

Example 5.5 Two small surfaces 1 and 2 are oriented perpendicularly to each other as shown in Fig. 5.13 and have surfaces 0.1 and 0.2 m^2, respectively, and

Table 5.3 Examples of formulas for calculating the view factors between finite areas

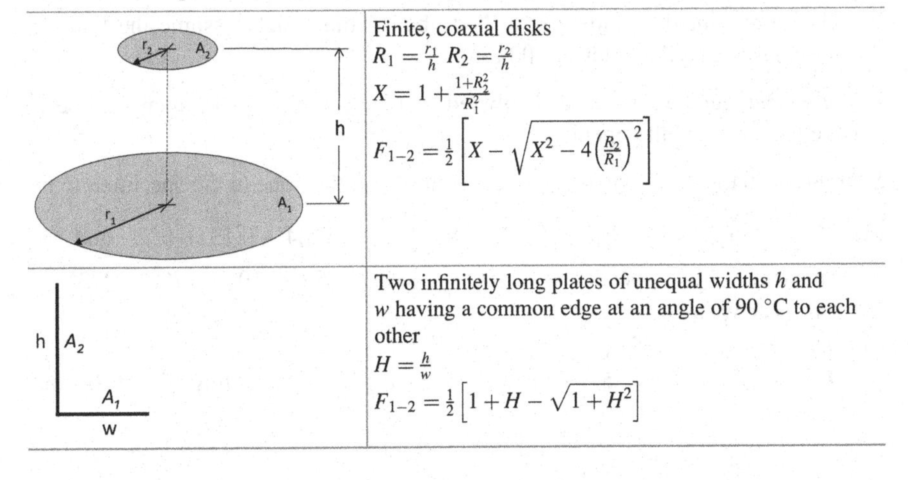

(figure)	Finite, coaxial disks $R_1 = \frac{r_1}{h}$ $R_2 = \frac{r_2}{h}$ $X = 1 + \frac{1+R_2^2}{R_1^2}$ $F_{1-2} = \frac{1}{2}\left[X - \sqrt{X^2 - 4\left(\frac{R_2}{R_1}\right)^2}\right]$
(figure)	Two infinitely long plates of unequal widths h and w having a common edge at an angle of 90 °C to each other $H = \frac{h}{w}$ $F_{1-2} = \frac{1}{2}\left[1 + H - \sqrt{1 + H^2}\right]$

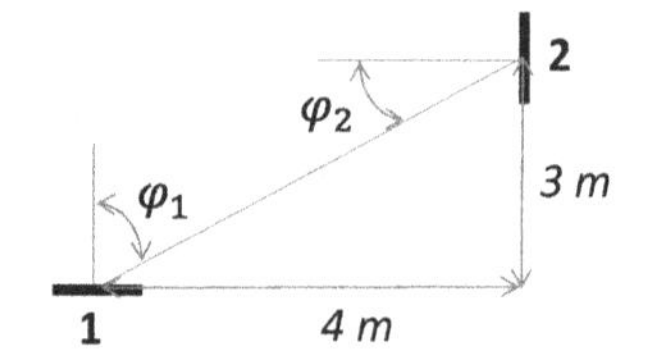

Fig. 5.13 Two small surfaces in the same plane oriented perpendicularly

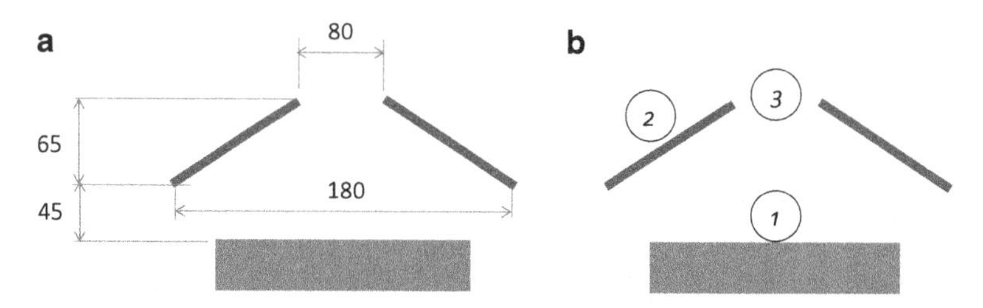

Fig. 5.14 Sketch of cone calorimeter for calculation of view factors to the specimen. The surfaces are identified by the numbers in circles. (**a**) Measures in mm. (**b**) Numbering of surfaces

temperatures 850 and 400 K, respectively. How much heat is transferred between the two surfaces?

Solution The distance between the surfaces becomes 5 m and thus $\cos\varphi_1 = 0.6$ and $\cos\varphi_2 = 0.8$. Then Eq. 5.33 yields $\dot{q}_{d1-d2} = 5.67 \cdot 10^{-8} \left(850^4 - 400^4\right) \frac{0.6 \cdot 0.8}{\pi\, 5^2} \cdot 0.1 \cdot 0.2 = 3.44$ W.

Example 5.6 In the Cone Calorimeter the radiant panel has the shape of a truncated circular cone as shown in Fig. 5.14. Assuming the panel has a uniform temperature of 700 °C and an emissivity equal unity, and neglecting the radiation from outside the cone, calculate

(a) The *maximum* incident radiation to a body below the panel.
(b) The *mean* incident radiation to body below the panel. Assume the body is circular with a diameter of 100 mm.

Guidance: Surfaces may be numbered as Fig. 5.14b, i.e. specimen surface is 1, cone heater 2 and the opening 3.

Solution Equation 5.28 yields the view factor from the cone to the specimen to be

(a) $F_{d1-2} = F_{d1-2,3} - F_{d1-3} = \frac{90^2}{40^2+90^2} - \frac{40^2}{105^2+40^2} = 0.84 - 0.13 = 0.71$ and the incident flux becomes $q_{inc,\ d1} = 0.71 \cdot 5.67 \cdot 10^{-8} \cdot (700 + 273)^4 = 36{,}000\ \text{W/m}^2$.
(b) $F_{1-2} = F_{1-2,3} - F_{1-3}$
$F_{1-2,3}$: From Table 5.3 with $R_1 = \frac{50}{40}$ and $R_2 = \frac{90}{40}$ yielding $X = 4.88$ and $F_{2,3-1} = 0.79$

F_{1-3}: $R_1 = \frac{50}{105}$ and $R_2 = \frac{40}{105}$ yielding $X = 6.05$ and $F_{1-3} = 0.11$. Thus $F_{1-2} = 0.79 - 0.11 = 0.68$ and Eq. 5.42 $q_{inc,\ 1} = 0.68 \cdot 5.67 \cdot 10^{-8} \cdot (700 + 273)^4 = 34{,}600\ \text{W/m}^2$. Comment: The mean incident flux is only by 4 % less than the maximum.

5.3 Radiation from Flames and Smoke

It is flames, smoke particles and combustion products that absorb and emit heat radiation in fires. It is generally assumed continuous over all wavelengths when calculating temperature although some gas species only absorb and emits at certain wavelength intervals. In general simple gas molecules such as oxygen O_2 and nitrogen N_2 do not absorb or emit heat radiation while molecules such as carbon monoxide CO_2 and water H_2O do depending on wavelength. Therefore the heat absorbed or emitted by clean air is negligible.

Overall the absorption α_{fl} and the emission ε_{fl} of a flame or smoke layer depend on the *absorption or emission coefficient* K and the *mean beam length* L_e. According to the Kirchhoff's law the absorptivity and the emissivity are equal. Then the Beer's law is a useful tool in approximate radiation analyses [1, 2]. Thus

$$\alpha_{fl} = \varepsilon_{fl} = 1 - e^{-K \cdot L_e} \tag{5.44}$$

For gas species K depends on wavelength, but as the bulk of the radiation from flames and smoke layers emanates from soot particles, it is treated as independent of wavelength, i.e. K is treated as an effective absorption/emission coefficient.

The emitted heat from a flame may accordingly be written as

$$\dot{q}''_{emi,fl} = \left(1 - e^{-K \cdot L_e}\right) \sigma \cdot T_{fl}^4 \tag{5.45}$$

where T_{fl} is the flame temperature (assumed uniform). A few empirical and not very reliable data for the effective absorption/emission coefficient, K, are available in the literature. Some values are shown in Table 5.4.

The mean beam length giving reasonable approximations may be obtained from

$$L_e = 3.6 \frac{V}{A} \tag{5.46}$$

where V is the total volume of the gas and A the total surface area. For a volume between two infinite planes at a distance L a mean beam length L_e can be obtained as

$$L_e = 1.8\, L \tag{5.47}$$

Table 5.4 The effective absorption/emission coefficient K for various fuels, from [13]

Fuel	K (m^{-1})	Reference
Diesel oil	0.43	Sato and Kunimoto
Polymethylmethacrylate	0.5	Yuen and Tien
Polystyrene	1.2	Yuen and Tien
Wood cribs	0.8	Hägglund and Persson
Wood cribs	0.51	Beyris et al.
Assorted furniture	1.1	Fang

Observe that a flame or a smoke layer absorbs radiant heat depending on the absorptivity according to Eq. 5.44. This is illustrated by Example 5.9.

Example 5.7 What is the emitted radiation heat flux $\dot{q}''$ from an oil fire where $K = 0.4\ m^{-1}$.

Assume a beam length $L = 1$ m and a flame temperature of $T_{fl} = 1073$ K (= 800 °C).

Solution

$$\dot{q}'' = \left(1 - e^{-K \cdot L}\right) \cdot \sigma \cdot T_{fl}^4 = 25{,}000\,\mathrm{W/m^2}$$

Example 5.8 The surface temperature T_s of a stove is 500 °C and has an emissivity of $\varepsilon_s = 1.0$. Near the stove is a wooden wall with a surface emissivity of $\varepsilon_w = 0.8$. The air temperature in the space between the stove and the wall $T_g = 40$ °C and the convection heat transfer coefficient is $h_c = 10\ \mathrm{W/(m^2 K)}$. Assume the surfaces of the stove and the wall being parallel and infinitely large.

(a) What is the net heat transfer by radiation to the wall surface at the ignition temperature assumed to be $T_{ig} = 300$?
(b) What is the maximum temperature the wall can obtain at equilibrium, when the surface does not absorb any more heat and is assumed to be a perfect insulator (i.e. the adiabatic surface temperature).

Solution

(a) Equations 5.11 and 5.12 yield $\dot{q}''_{rad} = \varepsilon_r \cdot \sigma\left[(T_s + 273)^4 - (T_w + 273)^4\right]$ and $\varepsilon_r = \frac{1}{\frac{1}{\varepsilon_s} + \frac{1}{\varepsilon_w} - 1 = \frac{1}{\frac{1}{1} + \frac{1}{0.8} - 1 = 0.8}}$, and $\dot{q}''_{rad} = 0.8 \cdot 5.67 \cdot 10^{-8} \cdot \left(773^4 - 573^4\right) = 11.3 \cdot 10^3\ \mathrm{W/m^2}$.

(b) The surface heat balance: $\varepsilon_r \sigma\left[(T_s + 273)^4 - (T_w + 273)^4\right] + h\left(T_g - T_w\right) = 0.8 \cdot 5.67 \cdot 10^{-8} \cdot \left[773^4 - (T_w + 273)^4\right] + 10 \cdot (40 - T_w) = 0$.

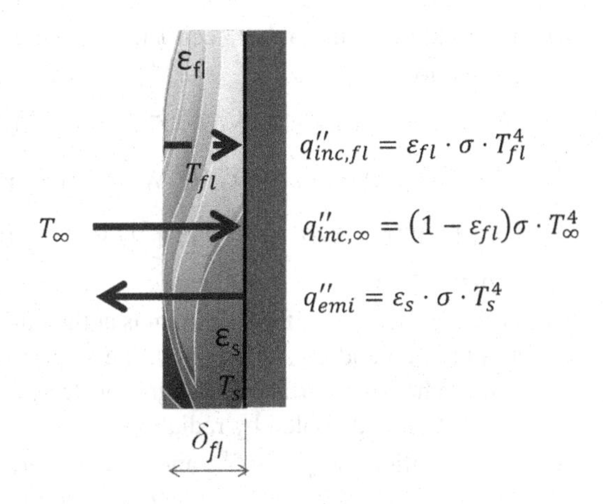

Fig. 5.15 One-dimensional model of the heat exchange by radiation at a surface exposed to a flame of limited thickness

An iteration scheme yields: $T_w^{i+1} = 273 + \left[773 - 10 \cdot \left(40 - T_w^i\right)/\left(0.8 \cdot 5.67 \cdot 10^{-8}\right)\right]^{1/4}$ °C. Assuming $T_w^0 = 300$ °C yields $T_w^1 = 467$ °C and subsequently $T_w^2 = 443$ °C and $T_w^3 = 446$ °C which is an acceptable solution.

Example 5.9 A specimen as shown in Fig. 5.15 is suddenly exposed to a propane flame which is assumed to have a thickness of $\delta_{fl} = 0.2$ m at the point being analysed. Assume the effective flame absorption coefficient $K = 0.5\ \mathrm{m}^{-1}$, the flame temperature $T_{fl} = 800$ °C, the ambient and initial temperatures $T_\infty = T_i = 20$ °C, the convective heat transfer coefficient $h_c = 10\ \mathrm{W/m^2}$, the surface emissivity $\varepsilon_s = 0.9$ and the ignition temperature of the specimen $T_{ig} = 350$ °C

(a) Calculate the incident radiant heat flux q''_{inc} to the specimen surface.
(b) Calculate the total heat flux by radiation and convection to specimen surface at the start of the test and at ignition, i.e. when the specimen surface temperature $T_s = 20$ °C and $T_s = 350$ °C, respectively.
(c) Comment on the magnitude of the contributions to the heat transfer by radiation and convection, respectively.
(d) Calculate the adiabatic surface temperature T_{AST} at the specimen surface.
(e) Repeat item (b), i.e. calculate the total heat flux to specimen at the start of the test and at ignition using T_{AST}.

Solution

(a) Incident heat flux from the flame, see Fig. 5.15, can be written as: $\dot{q}''_{inc} = \dot{q}''_{inc,fl} + \dot{q}''_{inc,\infty} = \varepsilon_{fl} \cdot \sigma \cdot T_{fl}^4 + (1 - \varepsilon_{fl})\sigma \cdot T_\infty^4$. The emissivity of a flame or smoke layer, ε_{fl}, may be calculated according to Eq. 5.45 as: $\varepsilon_{fl} = 1 - e^{-0.5 \cdot 1.8 \cdot 0.2} \approx 0.18$. The incident heat flux from the flame and the surrounding will be: $\dot{q}''_{inc} = 5.67 \cdot 10^{-8}\left[0.18\,(800 + 273)^4 + (1 - 0.18) \cdot (20 + 273)^4\right] = 13{,}871\ \mathrm{W/m^2} \approx 13.9\ \mathrm{kW/m^2}$.

(b) The total heat flux to the specimen is the sum of the heat flux by radiation and convection: $\dot{q}''_{tot} = \varepsilon_s(\dot{q}''_{inc} - \sigma T_s^4) + h_c(T_{fl} - T_s)$. At the initial temperature: $q''_{tot,20} = 0.9\left(13,871 - 5.67\ 10^{-8}\ (20+273)^4\right) + 10\ (800-20) = 12,108$ $+7800 = 19,908 \approx 19.9 \cdot 10^3\ \text{W/m}^2$ and at ignition temperature $q''_{tot,350} =$ $0.9\left(13,871 - 5.67\ 10^{-8}\ (350+273)^4\right) + 10\ (800-350) = 4796 + 4500$ $= 9.3 \cdot 10^3\ \text{W/m}^2$.

(c) When the surface of the specimen is at the initial temperature the contributions by radiation and convection is in the same order of magnitude in the studied case. When the surface is at ignition temperature, the surface is heated by convection and cooled by radiation.

(d) By definition $q''_{inc} = \sigma T_r^4$ and thus based on the q''_{inc} calculated above $T_r = 703\ \text{K} = 430\ °\text{C}$. T_{AST} can be obtained by solving the fourth degree equation 4.22. Thus $T_{AST} = 745\ \text{K} = 472\ °\text{C}$.

(e) Equation 4.31 yields at the initial temperature $\dot{q}''_{tot,20} = 0.9\,\sigma(703^4 - 293^4)$ $+10\,(703-293) = (15,343 + 4520) = 19.9 \cdot 10^3\ \text{W/m}^2$, and at the ignition temperature $\dot{q}''_{tot,350} = 0.9 \cdot \sigma(745^4 - 623^4) + 10 \cdot (745-623) = 9.3 \cdot 10^3\ \text{W/m}^2$. Note that these alternatively calculated heat flux values are equal to those calculated under item (b).

Example 5.10 A 6 m high, 4 m wide and 0.5 m thick flame is covering a well-insulated façade with a surface emissivity $\varepsilon_s = 0.9$. In the centre-line of the flame 2 m outside the façade surface is a small square section steel column. Assume the flame temperature equal $T_{fl} = 800\ °\text{C}$, the flame absorption/emission coefficient $\kappa = 0.3\ \text{m}^{-1}$ and the convective heat transfer coefficient $h = 35\ \text{W/(m}^2\ \text{K)}$, see Fig. 5.16.

Calculate under state conditions

(a) The temperature of the façade surface behind the flame.
(b) The emitted radiant flux from the flame surface towards the column.
(c) The maximum incident radiation to the four sides of the column.

Solution

(a) The emissivity according to Eqs. 5.44 and 5.47 $\varepsilon_{fl} = \left(1 - e^{-1.8*2*0.3}\right) = 0.66$. Then the incident radiation to the façade surface becomes $0.66 \cdot 5.67 \cdot 10^{-8}$ $(800+273)^4 = 49.6 \cdot 10^3\ \text{W/m}^2$ (corresponding to a radiation temperature of about $T_r = 700\ °\text{C}$) and the adiabatic surface temperature T_{AST} can then be obtained from (Eq. 4.21) as $0.9 \cdot \left[49.6 \cdot 103 - 5.67 \cdot 10^{-8} \cdot T_{AST}^4\right] +$ $35 \cdot (800 + 273 - T_{AST}) = 0$ which yields $T_{AST} = 988\ \text{K} = 715\ °\text{C}$ after two iterations according to Eq. 4.25.

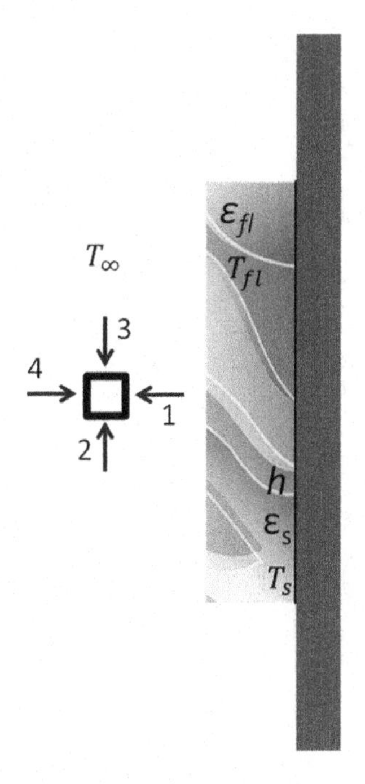

Fig. 5.16 Façade flame radiating on an external column surrounded by air at ambient temperature

(b) Contributions from the flame plus the façade surface $\dot{q}''_{emi} = 0.66 \cdot 5.67 \cdot 10^{-8} (800 + 273)^4 + (1 - 0.66) \cdot 5.67 \cdot 10^{-8} \cdot 988^4 = 68.0 \cdot 10^3\ \mathrm{W/m^2}$.

(c) *Column side facing the façade, 1*: Divide into four equal contributions according to Table 5.2 first row or the diagram in Fig. 5.9 with $X = 2/2 = 1$ and $Y = 3/2 = 1.5$ yields $F_{d1\text{-}2} = 0.09$ and the total view factor $F = 4 \cdot 0.09 = 0.36$ and thus incident flux $0.36 \cdot 68.0 \cdot 10^3 = 24.5 \cdot 10^3\ \mathrm{W/m^2}$.

Sides of the column, 2 and 3: The sides will only be exposed to half the flame. Figure 5.10 with $X = 2/2 = 1$ and $Y = 3/2 = 1.5$ yields $F_{d1\text{-}2} = 0.065$ and $F = 2 \cdot 0.065 = 0.15$ and thus incident flux $0.15 \cdot 68.0 \cdot 10^3 = 10.2 \cdot 10^3\ \mathrm{W/m^2}$.

Column side facing away from the façade, 4: This side does not face the façade and will only get an incident radiation corresponding to the ambient temperature, i.e. $5.67 \cdot 10^{-8}(20 + 273)^4 = 0.42 \cdot 10^3\ \mathrm{W/m^2}$.

Chapter 6
Heat Transfer by Convection

In previous chapters heat transfer by convection or just convection was treated only to the extent that it provides a linear boundary condition of the 3rd kind for conduction problems when the heat transfer coefficient is assumed constant. In this chapter the physical phenomenon of convection is described in more detail.

Heat is transferred by convection from a fluid to a surface of a solid when they have different temperatures. Here it is shown how the convection can be calculated and in particular how the *convection heat transfer coefficient,* denoted h or sometimes for clarity h_c, can be estimated in various situations relevant for FSE problems.

When the gas or liquid flow is induced by a fan, etc. it is called *forced convection*, and when it is induced by temperature differences between a surface and the adjacent gases it is called *natural convection* or *free convection*. In the latter case the surface heats or cools the fluid which then due to buoyancy moves upwards or downwards. Both natural and forced convection can be *laminar* or *turbulent* depending on fluid properties and velocity, and on size and shape of exposed surfaces. Various modes occur in fires and are relevant in FSE.

The heat transfer by convection depends in any case on the temperature difference between the fluid and the surface. Usually in FSE it is assumed directly proportional to the difference of the two temperatures according to the Newton's law of cooling, see Sect. 4.2. This is linear boundary condition which facilitates calculations without jeopardizing accuracy as heat transfer by radiation at elevated temperatures dominates over the transfer by convection.

Section 6.1 gives expressions on how air and water conductivity and viscosity vary with temperature. Viscosity is the measure of a fluid's resistance to flow and has a decisive influences convective heat transfer properties.

In Sects. 6.2 and 6.3 general formulas are presented for various fluids, configurations and flow conditions followed by some useful approximate formulas and diagrams applicable specifically to air which considerably facilitates calculations of FSE problems.

U. Wickström, *Temperature Calculation in Fire Safety Engineering*,
DOI 10.1007/978-3-319-30172-3_6

6.1 Heat Transfer Properties of Air and Water

The properties of several fluids are tabulated in textbooks such as [1] for various temperature levels. Special attention is given to air as it usually in flow calculations is assumed to have the same properties as smoke and fire gases.

For *air* the *conductivity* k_{air} can be approximated as [14]

$$k_{air} = 291 \cdot 10^{-6}\, T^{0.79}\, [\mathrm{W/(m\,K)}] \tag{6.1}$$

and the *kinematic viscosity* ν_{air} as

$$\nu_{air} = 1.10 \cdot 10^{-9}\, T^{1.68}\, \left[\mathrm{m^2/s}\right] \tag{6.2}$$

respectively, where T is temperature in Kelvin. See also Figs. 6.1 and 6.2 for graphical presentations. The simple approximations are used in this book for obtaining close form expressions for among other things convection heat transfer coefficients.

The *Prandtl number* does not vary much with temperature and may in most cases be assumed constant, $Pr_{air} = 0.7$.

The thermal conductivity *water* k_w can be approximated as

$$k_w = -0.575 + 6.40 \cdot 10^{-3} \cdot T - 8.2 \cdot 10^{-6} \cdot T^2\, [\mathrm{W/(m\,K)}] \tag{6.3}$$

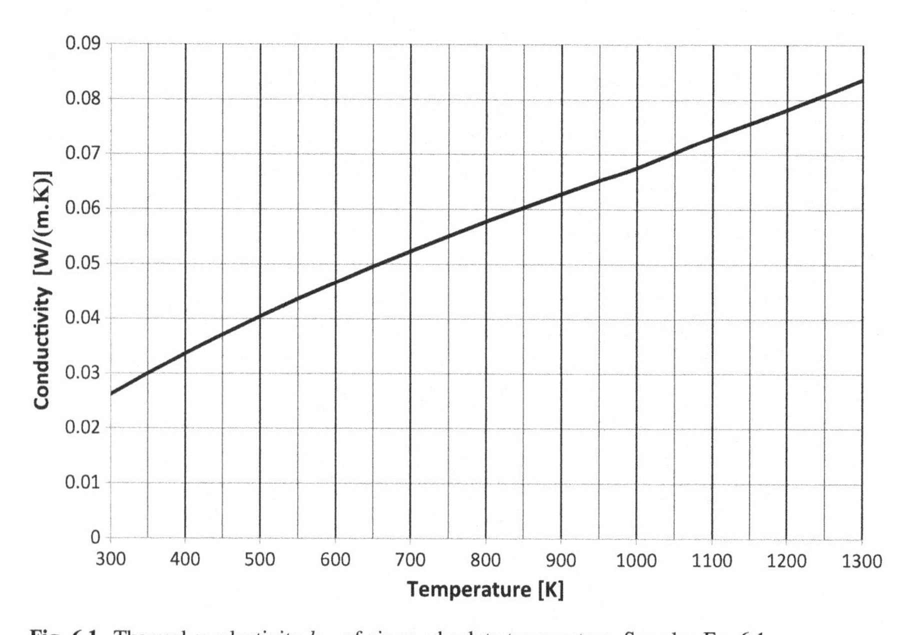

Fig. 6.1 Thermal conductivity k_{air} of air vs. absolute temperature. See also Eq. 6.1

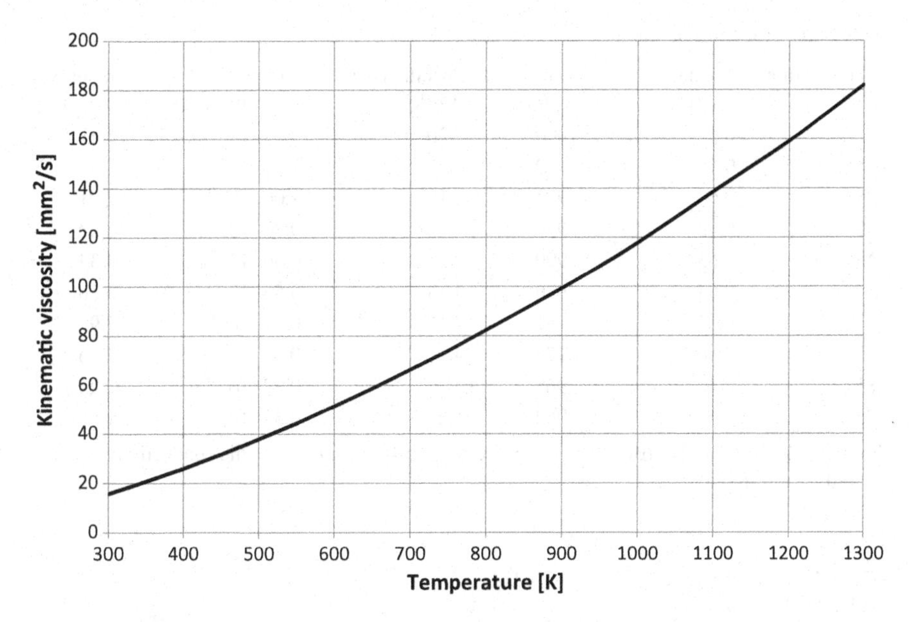

Fig. 6.2 Kinematic viscosity ν_{air} of air vs. absolute temperature. See also Eq. 6.2

where T is the water temperature in Kelvin. Table 6.1 shows also other properties of water at various temperatures relevant for thermal calculations.

Latent heat of vaporization of water relevant when calculating temperature in moist solids $a_w = 2260\,\mathrm{kJ/kg}$.

6.2 Forced Convection

The heat transfer coefficient or the thermal resistance between a gas or fluid and a solid surface is controlled within the so-called boundary layer. The thermal resistance of this layer and thereby the amount of heat being transferred depends on the thickness of the layer and the conductivity of the fluid. The thickness of the boundary layer in turn depends on the velocity of the fluid.

6.2.1 On Flat Plates

Figure 6.3 shows an edge where a boundary layer is developed in a forced flow over a *flat surface*. Outside the boundary layer the flow is undisturbed and has a uniform velocity u_∞. It then decreases gradually closer to the surface and very near the surface it vanishes. As indicated in Fig. 6.3 the boundary layer thickness δ grows with the distance from the leading edge.

Table 6.1 Properties of water

Temperature (°C)	Conductivity (W/(m K)	Density (kg/m^3)	Specific heat (kJ/(kg K))	Kinematic viscosity (m^2/s)	Prandtl's no. (–)
5	0.57	1000	4.20	$1.79 \cdot 10^{-6}$	13.67
15	0.59	999	4.19	$1.30 \cdot 10^{-6}$	9.47
25	0.60	997	4.18	$1.00 \cdot 10^{-6}$	7.01
35	0.62	994	4.18	$0.80 \cdot 10^{-6}$	5.43
45	0.63	990	4.18	$0.66 \cdot 10^{-6}$	4.34
55	0.64	986	4.18	$0.55 \cdot 10^{-6}$	3.56
65	0.65	980	4.19	$0.47 \cdot 10^{-6}$	2.99
75	0.66	975	4.19	$0.41 \cdot 10^{-6}$	2.56
85	0.67	968	4.20	$0.37 \cdot 10^{-6}$	2.23
95	0.67	962	4.21	$0.33 \cdot 10^{-6}$	1.96

From the Engineering ToolBox, www.EngineeringToolBox.com, except the conductivity which is according to Eq. 6.3.

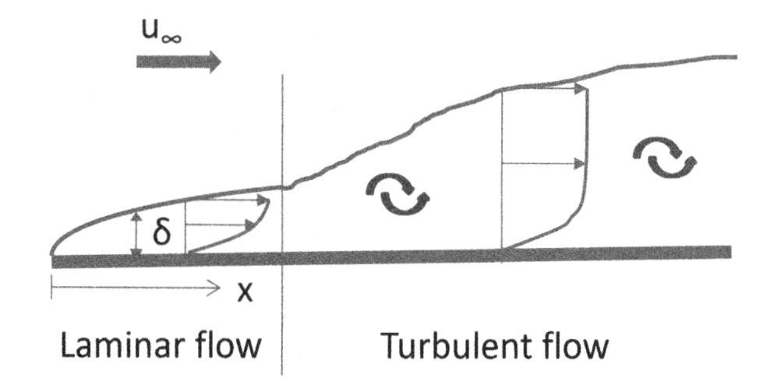

Fig. 6.3 Boundary layer with a thickness δ developing after an edge of a flat surface

The *heat transfer resistance* R_h can now be calculated as heat resistance between the fluid and the solid surface in a similar way as for thermal conduction in solids:

$$R_h = \frac{\delta}{k_f} \tag{6.4}$$

where k_f is the thermal conductivity of the fluid and δ the boundary layer thickness. The subscript f indicates that the parameter values shall be at the film temperature which is the average of the surface T_s and the free stream fluid T_∞ temperatures, i.e.

$$T_f = \frac{(T_s + T_\infty)}{2} \tag{6.5}$$

The heat transfer coefficient can be obtained as the inverse of the heat transfer resistance according to Eq. 2.14. Thus

$$h_f = \frac{k_f}{\delta} \tag{6.6}$$

The heat transfer coefficient is often expressed by the Nusselt number Nu, a non-dimensional relation between the boundary layer thickness δ and a characteristic length x of the exposed surface. With δ obtained from Eq. 6.6

$$Nu_f = \frac{x}{\delta} = \frac{h \cdot x}{k_f} \tag{6.7}$$

In the case of a plane surface as shown in Fig. 6.3, the characteristic length x is the distance from the edge. Near the edge, small values of x, the flow is laminar and further away it is turbulent. The Nusselt number at a distance x has been derived analytically as (see, e.g. [1]):

$$Nu_{xf} = 0.332\, Pr_f^{1/3} \cdot Re_{fx}^{1/2} \tag{6.8}$$

where Pr_f is the Prandtl number which relates the *kinematic viscosity* ν and *thermal diffusivity* α of the fluid. Re_{fx} is the *Reynolds number* which indicates whether the flow conditions are laminar or turbulent. It is a non-dimensional grouping of parameters defined as

$$Re_{fx} = \frac{u_\infty \cdot x}{\nu_f} = \frac{\rho_f \cdot u_\infty \cdot x}{\mu_f} \tag{6.9}$$

where μ is the *dynamic viscosity* of the fluid. The kinematic viscosity ν is the dynamic viscosity divided by the density, i.e.:

$$\nu = \frac{\mu}{\rho} \tag{6.10}$$

By integration along the surface the mean Nusselt number $\overline{Nu_f}$ can be obtained as

$$\overline{Nu_f} = 2 \cdot Nu_{xf} = 0.664 \cdot Pr_f^{1/3} \cdot Re_{fx}^{1/2} \tag{6.11}$$

For constant fluid properties the heat transfer coefficient at a distance x can now be calculated by combining Eqs. 6.7 and 6.8:

$$h_{fx} = \frac{Nu_{xf} \cdot k_f}{x} = 0.332 \cdot k_f \cdot Pr_f^{\frac{1}{3}} \cdot \nu_f^{-1/2} \cdot u_\infty^{1/2} \cdot x^{-1/2} \tag{6.12}$$

The mean heat transfer coefficient from the edge $x = 0$ to x is twice this value:

$$\overline{h_{fx}} = 2\, h_{fx} = 0.664\, k_f \cdot Pr_f^{\frac{1}{3}} \cdot \nu_f^{-1/2} u_\infty{\cdot}^{1/2} x^{-1/2} \tag{6.13}$$

An important observation is that the heat transfer coefficient decreases with dimensions. Smaller dimensions mean larger convective heat transfer coefficients.

6.2.1.1 Heat Transfer in Air

To facilitate analyses of FSE problems material properties of air is now assumed by inserting Eqs. 6.1 and 6.2 into Eq. 6.12 and assuming Pr = 0.7. The heat transfer coefficient can then be obtained as a function of film temperature, air velocity and distance from the edge as:

$$h_{fx} = 2.59\ \cdot T_f^{-0.045} \cdot u_\infty{}^{1/2} \cdot x^{-1/2}\ \left[\mathrm{W}/\left(\mathrm{m}^2\,\mathrm{K}\right)\right] \tag{6.14}$$

Note that the influence of the film temperature level is rather weak. The mean heat transfer coefficient will be twice that value:

$$\overline{h_{fx}} = 2 \cdot h_{fx} = 5.17 \cdot\ T_f^{-0.045} \cdot u_\infty{}^{1/2} \cdot x^{-1/2}\ \left[\mathrm{W}/\left(\mathrm{m}^2\,\mathrm{K}\right)\right] \tag{6.15}$$

Example 6.1 A 200-mm-wide steel plate having a uniform temperature of 500 K is exposed to an air stream with a temperature of 1200 K and a velocity of 2.0 m/s. Calculate the mean heat flux by convection $\dot{q}''_{\mathrm{con}}$ to the steel surface.

Solution Insert the parameters in Eq. 6.15. The film temperature $T_f = 850\, K$ and then $\overline{h_{fx}} = 5.17 \cdot\ 850^{-0.045} \cdot 2^{\frac{1}{2}} \cdot 0.2^{-\frac{1}{2}}\ \mathrm{W}/(\mathrm{m}^2\ \mathrm{K}) = 12.0\,\mathrm{W}/(\mathrm{m}^2\ \mathrm{K})$, and the mean heat flux $\dot{q}''_{\mathrm{con}} = 12.0 \cdot (1200 - 500) = 8370\ \mathrm{W/m}^2$.

Example 6.2 A heat flux meter measures the heat flux to its cooled sensor surface, see Sect. 9.2. Assuming the sensor surface has a diameter of 10 mm, calculate the heat transfer coefficient by convection $\bar{h}$. The sensor surface temperature is estimated to be uniform and equal to 30 °C. Estimate the heat flux by convection $\dot{q}''_{\mathrm{con}}$ to the sensor surface if the gas temperature is 400 °C and the gas velocity is 2 m/s.

Solution Apply Eq. 6.15 $\bar{h} = 5.17 \cdot \left(\frac{400+30}{2} + 273\right)^{-0.045} \cdot 2^{1/2} \cdot 0.01^{-1/2} = 55\,\mathrm{W/m}^2$, and the heat flux to the sensor surface by convection becomes $\dot{q}''_{\mathrm{con}} = 55 \cdot (400 - 30) = 20300\,\mathrm{W/m}^2 = 20.3\,\mathrm{kW/m}^2$.

6.2.2 *Across Cylinders*

In the case of flow across a *cylinder* an empirical expression for the Nusselt number has been derived, see, e.g. [1, 2]:

$$Nu_f = \frac{h \cdot d}{k_f} = C \cdot \left(Re_{df}\right)^n \cdot Pr_f^{\frac{1}{3}} \tag{6.16}$$

For a cylinder with a diameter d and the Reynolds number becomes

$$Re_{df} = \frac{u_\infty \cdot d}{\nu_f} = \frac{u_\infty \cdot d}{1.13 \cdot 10^{-9} \cdot T_f^{1.67}} = \frac{1}{1.13 \cdot 10^{-9}} \cdot \Gamma \tag{6.17}$$

where the parameter group named Γ identified as

$$\Gamma = \frac{u_\infty \cdot d}{T_f^{1.67}} \tag{6.18}$$

is introduced to simplify calculations.

The convection heat transfer coefficient can be derived from Eq. 6.16 as

$$h = \frac{k_f}{d} \cdot C \cdot \left(\frac{u_\infty \cdot d}{\nu_f}\right)^n \cdot Pr_f^{\frac{1}{3}} \tag{6.19}$$

The constants C and n are given in Table 6.2.

6.2.2.1 Heat Transfer in Air

For air or fire gases the conductivity k_f and the viscosity ν_f as functions of temperature can be obtained from Eqs. 6.1 and 6.2, respectively, and inserted into Eq. 6.19 to become

$$h = \frac{291 \cdot 10^{-6} \cdot T_f^{0.79}}{d} \cdot C \cdot \left(\frac{u_\infty \cdot d}{1.10 \cdot 10^{-9} \cdot T_f^{1.68}}\right)^n \cdot 0.7^{\frac{1}{3}} \tag{6.20}$$

which can be reduced to

$$h = A \cdot C \cdot \frac{T_f^{0.79-1.68n} \cdot u_\infty{}^n}{d^{1-\mathrm{n}}} \tag{6.21}$$

where the constants A, C and n can be found in Table 6.2 for various ranges of the values of Re_{df} or Γ.

Table 6.2 Constants to be used with Eqs. 6.16 and 6.21 for calculating the Nusselt number and heat transfer coefficients to cylinders exposed to forced convection flow

Re_{df}	Γ (Eq. **6.18**)	C	n	A [SI units]
0.4–4	$(0.45–4.5) \cdot 10^{-9}$	0.989	0.330	0.110
4–40	$(4.5–45) \cdot 10^{-9}$	0.911	0.385	0.341
40–4000	$(45–4500) \cdot 10^{-9}$	0.683	0.466	1.81
4000–40,000	$(4.5–45) \cdot 10^{-6}$	0.193	0.618	41.4
40,000–400,000	$(45–450) \cdot 10^{-6}$	0.0266	0.805	1950

Example 6.3 Calculate the convective heat transfer coefficient of a 1 mm shielded thermocouple. Model the thermocouple as a cylinder and assume an air velocity of $u_\infty = 1.0\,\mathrm{m/s}$ flowing across the cylinder. Assume gas temperature levels of

(a) $T_g = 300$ K (room temperature initially)
(b) $T_g = 1000$ K (ultimate temperature)

Solution

(a) The film temperature $T_f = (300 + 1000)/2 = 650$ K. Then according to Eq. 6.18 $\Gamma = 1 \cdot 0.001/\left(650^{1.67}\right) = 20 \cdot 10^{-9}$ and from Table 6.2 $C = 0.911$, $n = 0.385$ and $A = 341$. Then $h = 0.341 \cdot 0.911 \cdot \frac{T_f^{0.79-1.68 \cdot 0.385} u_\infty{}^{0.385}}{d^{1-0.385}} =$ 131 $\mathrm{W/(m^2K)}$.
(b) Equation 6.18 yields $\Gamma = 1.0 \cdot 0.001/\left(1000^{5/3}\right) = 9.77 \cdot 10^{-9}$. A, C and n can now be obtained from row two of Table 6.2 and inserted into Eq. 6.21 to get $h = 0.341 \cdot 0.911 \cdot \frac{T_f^{0.92-1.67 \cdot 0.385} u_\infty{}^{0.385}}{d^{1-0.385}} = 0.311 \cdot \frac{T_f^{0.277} u_\infty{}^{0.385}}{d^{0.615}} = 147\mathrm{W/(m^2K)}$

Comment: The heat transfer coefficient due to convection changes only slightly with temperature. However, it would change considerably if the heat transfer due to radiation would be included as well.

6.2.3 *In Circular Pipes and Tubes*

The heat transfer between the fluid and the walls of a circular tube depends on the fluid conductivity k and kinematic viscosity ν, the fluid velocity u and the flow conditions, laminar or turbulent governed by the Reynolds number

$$Re_D = \frac{u \cdot D}{\nu} = \frac{\rho \cdot u \cdot D}{\mu} \tag{6.22}$$

All parameters refer to bulk temperatures. For details see textbooks such as [1, 2].

For *turbulent* flow $(Re_D > 2300)$ the Nusselt number can be calculated as

$$Nu_d = 0.023 \cdot Re_d^{0.8} \cdot Pr^n \tag{6.23}$$

where $n = 0.4$ for heating and $n = 0.3$ for cooling of the fluid.

For *laminar* flow $(Re_D < 2300)$ the Nu_d approaches a constant value for sufficiently long tubes, i.e.

$$Nu_d = 3.66 \tag{6.24}$$

Then the heat transfer coefficient between the fluid and the walls can be calculated as

$$h = Nu_d \cdot \frac{k_f}{D}. \tag{6.25}$$

where k_f is the conductivity of the fluid at the film temperature.

Example 6.4 Air with a bulk temperature $T_{air} = 200\,°\text{C}$ is flowing with a velocity of 2 m/s in a tube/duct with an inner diameter $D = 400$ mm. Estimate the heat flux to the duct surfaces which have a temperature of 800 °C?

Solution The film temperature $T_f = 0.5(200 + 800) = 500\ °\text{C} = 773$ K. Then Eq. 6.1 yields $k_{air} = 0.0567\,\text{W}/(\text{m K})$ and Eq. 6.2 $\nu_{air} = 78.3 \cdot 10^{-6}\ \text{m}^2/\text{s}$ and according to Eq. 6.22 $Re = 10{,}200$ which indicates turbulent flow. Then from Eq. 6.23 $Nu_d = 32.2$ (the fluid is heated and $n = 0.4$) and according to Eq. 6.25 $h = Nu_d \cdot \frac{k}{D} = 32.2 \cdot \frac{0.0567}{0.4} = 4.48\,\text{W}/(\text{m}^2\,\text{K})$. Thus the heat flux to the tube wall $\dot{q}''_w = 4.48 \cdot (800 - 200) = 2690\ \text{W}/\text{m}^2$.

Example 6.5 Water with a bulk temperature of $T_w = 20$ °C is flowing with a velocity $u = 0.1$ m/s in a tube with an inner diameter $D = 50$ mm. Estimate the heat flux to the tube surfaces which have a constant temperature $T_w = 70$ °C?

Solution The film temperature $T_w = \frac{20+70}{2} = 45$ °C $= 318$ K. Then from Table 6.1 the conductivity $k_w = 0.63\ \text{W}/(\text{m K})$, the viscosity $\nu_w = 0.66 \cdot 10^{-6}\ \text{m}^2/\text{s}$ and $Pr_w = 4.34$. According to Eq. 6.22 $Re = 7576$ which indicates turbulent flow. Then from Eq. 6.23 $Nu_d = 52.2$ (the fluid is heated and $n = 0.4$) and according to Eq. 6.25 $h = Nu_d \cdot \frac{k}{D} = 52.2 \cdot \frac{0.63}{0.4} = 658\ \text{W}/(\text{m}^2\,\text{K})$.

6.3 Natural or Free Convection

6.3.1 On Vertical and Horizontal Plates

Natural or free convection occurs as a result of density changes due to heating or cooling of fluids at solid surfaces. When a wall is hotter than adjacent air an

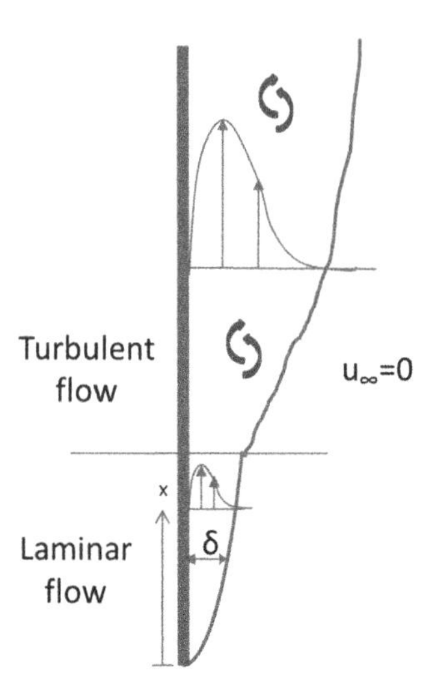

Fig. 6.4 Free convection boundary layer at a vertical hot surface

upwards flow is established as shown in Fig. 6.4, and vice versa if the wall is cooler. The velocity just at the wall surface is zero. It increases then to a maximum and thereafter it decreases to zero again at the end of the boundary layer where the free stream velocity is assumed to be negligible. At the outset the boundary layer is laminar but changes to turbulent at some distance from the edge depending on fluid properties and the difference between wall surface and fluid temperatures. In practice in FSE, natural and forced convection commonly occur simultaneously and analyses must focus on the one which is predominant.

In general it is very difficult to make accurate estimates of natural convection heat transfer coefficients. The formula given below for some elementary cases are based on empirical evidence obtained under controlled conditions. Such conditions rarely occur in real life but they serve as guidance for estimates. Any accurate analytical solutions are not available for calculating heat transfer by natural convection.

As for forced convection the Nusselt number yields the heat transfer coefficient. It depends in the case of free convection on the Prandtl and Grashof numbers. The latter is defined as

$$Gr_{xf} = \frac{g \cdot \beta_f \cdot (T_s - T_g) \cdot L^3}{v_f^2} \tag{6.26}$$

where g (=9.81 m/s^2) is the constant of gravity, L a characteristic length, ν the kinematic viscosity of the fluid and β_f is the inverse of the film temperature as defined in Eq. 6.5:

$$\beta_f = \frac{1}{T_f} \tag{6.27}$$

The subscript f indicates that values of the parameters are taken at the film temperature.

Based on empirical data for both laminar and turbulent convection the average values of the Nusselt number and the corresponding heat transfer coefficient can be written as:

$$\overline{Nu_f} = \frac{\overline{h_f} \cdot L}{k_f} = C \cdot \left(Gr_f \cdot Pr_f\right)^m \tag{6.28}$$

This formula correlates well for several simple geometrical configurations. Here L is a characteristic length appropriate for the configuration. For horizontal rectangular plates the *characteristic length* may be estimated as

$$L = \frac{A}{P} \tag{6.29}$$

where A is the area and P the perimeter of the surface. The other constants of Eq. 6.28 can be found in Table 6.3. As a rule of thumb, the exponent $m = ¼$ for laminar and $m = 1/3$ for turbulent flow.

6.3.1.1 Explicit Expressions for Heat Transfer from Air

Equation 6.28 and Table 6.3 can be used to derive convection heat transfer coefficients. By inserting the approximations of k_f (Eq. 6.1) and υ_f (Eq. 6.2) of air as functions of temperature, explicit expressions can be derived for heat transfer coefficients as functions of the air T_{air} and the surface T_s temperature. T_f is the film temperature defined as the mean of the ambient air temperature and the surface temperature (Eq. 6.5).

Observe temperatures must be in Kelvin in all formulas.

Thus at *vertical plates* and *large cylinders* under *turbulent* conditions ($m = 1/3$) in air the mean heat transfer coefficient can be calculated as

$$\overline{h_f} = \overline{Nu_f} \cdot \frac{k_f}{L} = 76.0 \cdot T_f^{-0.66} \cdot |T_s - T_{air}|^{1/3} \tag{6.30}$$

The heat flux to a vertical surface may now be written in the form given in Eq. 4.8 as

$$\dot{q}''_{\text{con}} = 76.0\, T_f^{-0.66} (T_{\text{air}} - T_{\text{s}})^{0.33} \tag{6.31}$$

Table 6.3 Constants to be used with Eq. 6.28 for calculating heat transfer coefficients and heat transfer to surfaces exposed to natural convection

Configurations	$Gr_{Lf}Pr_f$.	Characteristic length	C	m
Vertical plates and large cylinders				
– Laminar	10^4–10^9	L	0.59	1/4
– Turbulent	10^9–10^{12}	L	0.13	1/3
Horizontal plates				
– Laminar (heated surface up or cooled down)	$2 \cdot 10^4$–$8 \cdot 10^6$	$L = A/P$	0.54	1/4
– Turbulent (heated surface up or cooled down)	$8 \cdot 10^6$–0^{11}	$L = A/P$	0.14	1/3
– Laminar (heated surface down or cooled up)	10^5–10^{11}	$L = A/P$	0.27	1/4

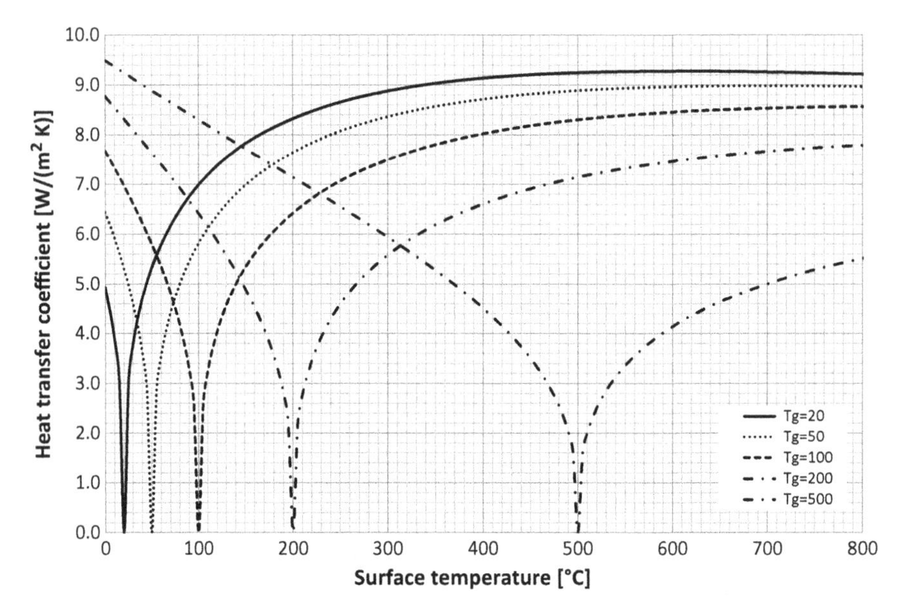

Fig. 6.5 Heat transfer coefficient due to natural convection vs. surface temperature for various surrounding air temperatures at ***vertical*** *plates* and *large cylinders* under *turbulent* conditions according to Eq. 6.30

$\dot{q}''_{con}$ is positive when $(T_{air} - T_s)$ is positive and vice versa. Notice that in this case the heat transfer coefficient is independent of the dimension L and inversely proportional to approximately the square root of the temperature level. The variation of h_f with the surface temperature for various gas temperatures are shown in Fig. 6.5. When the surface and gas temperatures are equal the air flow and thereby the heat transfer vanishes and it increases gradually when the fluid and surface temperatures diverges. As the formula are symmetric concerning T_{air} and T_s the diagram also be interpreted as $\overline{h_f}$ vs. gas temperature for various surface temperatures. Observe that for the gas temperature equal to 20 °C that convection heat transfer coefficient peaks at about 11.5 W/(m^2 K). Then it decreases slowly. In the interval from room temperature to 200 °C which is of interest for evaluation of fire separating walls the convection heat transfer coefficient increases from zero to about 11 W/(m^2 K).

Observe that Eqs. 6.30 and 6.31 apply only for turbulent conditions according to Table 6.3. This requirement is generally met in fire safety problems such as cooling of the unexposed side of a fire separating wall element.

The convective heat transfer coefficient to horizontal surfaces depends on the size of the surface. By inserting the value of the kinematic viscosity at the film temperature according to Eq. 6.2 the Rayleigh number becomes

$$\left(Pr_f \cdot Gr_{Lf}\right) = 5.68 \cdot 10^{18}\, L^3 \cdot T_f^{-4.36}\, \Delta T \tag{6.32}$$

If the Rayleigh number is between $2 \cdot 10^4$ and $8 \cdot 10^6$ according to Table 6.3, the Nusselt number can be obtained according to Eq. 6.28 with the air conductivity according to Eq. 6.1 and $Pr = 0.7$. The heat transfer coefficient can then be calculated as

$$\overline{h_f} = 7.67\, L^{-1/4} \cdot T_f^{-0.33} \cdot \Delta T^{1/4} \tag{6.33}$$

Of special interest are the heat transfer coefficients to specimen surfaces of the cone calorimeter (ISO 5660) and to plate thermometers when mounted horizontally. Both have a surface 0.1 m by 0.1 m and thus a characteristic length $L = 0.025$ m according to Eq. 6.29. Then the convection heat transfer coefficient can be calculated as

$$\overline{h_f} = 19.3\, T_f^{-0.33} \cdot \Delta T^{1/4} \tag{6.34}$$

Figure 6.6 shows convective heat transfer coefficients to horizontal surfaces with a characteristic length of 0.025 m for various gas temperatures as functions of the surface temperature.

Example 6.6 A PT is exposed to an incident radiation $\dot{q}''_{inc} = 50\ \mathrm{kW/m^2}$. Assuming that it does not lose any heat by conduction estimate its steady-state temperature, i.e. adiabatic surface temperature, when the PT is mounted

(a) horizontally
(b) vertically

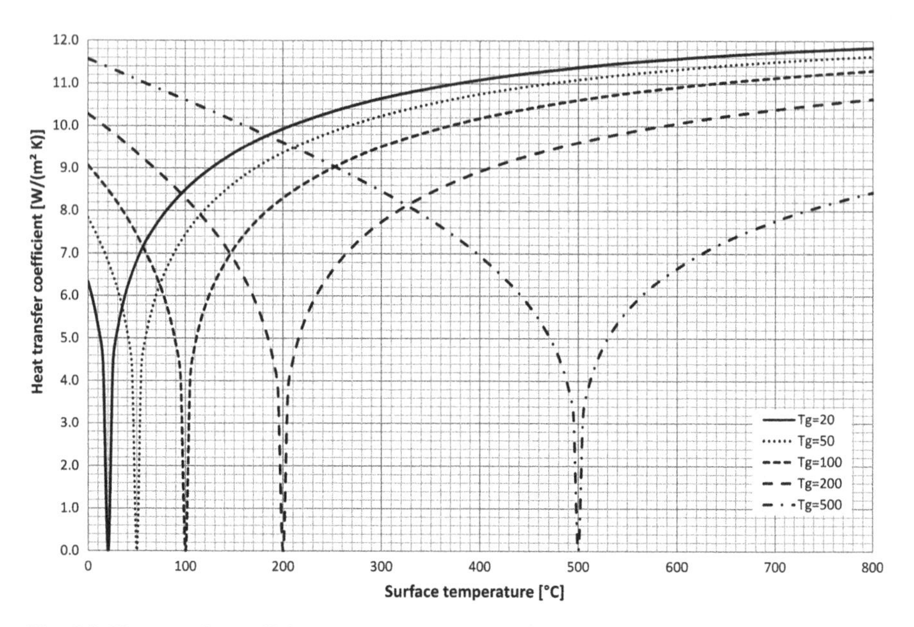

Fig. 6.6 Heat transfer coefficient vs. surface temperature due to natural convection at *a horizontal plate* for $L = 0.025$ m and $2 \cdot 10^4 < Gr_{Lf}Pr_f < 8 \cdot 10^6$. Particularly applicable to cone calorimeter (ISO 5660) and PTs with exposed areas of 0.1 m by 0.1 m

The exposed surface area of the PT is 0.1 m by 0.1 m. Assume the emissivity of the PT surface is 0.9 and the ambient temperature is 20 °C.

Hint: The net heat absorbed by radiation must be balanced by the heat lost by convection.

Solution $\dot{q}''_{inc} = 50\,\text{kW/m}^2$ yields a radiation temperature $T_r = 969$ K $= 696$ °C.

(a) Assuming a first surface temperature estimate $T^1_{AST} = T_s = T_r$ yields according to Eq. 6.34 or Fig. 6.6 $h = 11.7$ W/(m K). Then the AST can be calculated with the iteration procedure according to Eq. 4.25 with $T^1_{AST} = T_r$. Thus $T^2_{AST} = 923\,\text{K}$ and $T^3_{AST} = 926\,\text{K} = 653\,°\text{C}$. Finish iteration. Compare with Fig. 4.7a for $h/\varepsilon = 13\,\text{W/(m}^2\,\text{K)}$.

(b) Assuming a first surface temperature estimate $T^1_{AST} = T_s = T_r$ yields according to Eq. 6.30 or Fig. 6.5 $h = 9.2$ W/(m K). Then $T^2_{AST} = 934$ K and $T^3_{AST} = 936$ K $= 663$ °C. Finish iteration. Compare with Fig. 4.7a for $h/\varepsilon = 10\,\text{W/(m}^2\,\text{K)}$.

6.3.2 In Enclosed Spaces

Heat is transferred between surfaces of enclosures. Two elementary cases of free convection may be identified characterized by horizontal and vertical layers, respectively.

The heat flux may in both cases be calculated as [1, 15]

$$\dot{q}''_{c} = Nu_{\delta} \frac{k_f}{\delta} (T_1 - T_2) \tag{6.35}$$

where k_f is the thermal conductivity of the fluid. For horizontal layers where the upper surface is warmer there will be no buoyancy driven convection or flow and the heat will be transferred by conduction only, i.e. $Nu_{\delta} = 1$. However, if the upper surface is cooler than the lower convection will occur and the Nusselt number will be greater than one. Equation 6.35 may also be written as

$$\dot{q}''_{c} = \frac{k_e}{\delta} (T_1 - T_2) \tag{6.36}$$

where the k_e may be identified as the *effective* or *apparent thermal conductivity* of the air enhanced by convection. It is defined by the relation

$$\frac{k_e}{k_f} = Nu_{\delta} \tag{6.37}$$

The Nusselt numbers can be obtained from Table 6.3 for various ranges of the Grashof number according to Eq. 6.38

$$Gr_{\delta} = \frac{g \cdot \beta (T_1 - T_2)\, \delta^3}{\upsilon^2} \tag{6.38}$$

When the lower surface is warmer than the upper, convection and heat transfer by convection will occur when $Gr > 10^4$. Inserting the values of υ for *air* and β as functions of the mean of T_1 and T_2 according to Eqs. 6.2 and 6.27, respectively, yields:

$$Gr_{\delta}^{\text{air}} = \frac{8.11 \cdot 10^{18}\, (T_1 - T_2)\, \delta^3}{{T_{\delta}}^{13/3}} \tag{6.39}$$

where T_{δ} is the mean of the surface temperatures, i.e.

$$T_{\delta} = \frac{(T_1 + T_2)}{2} \tag{6.40}$$

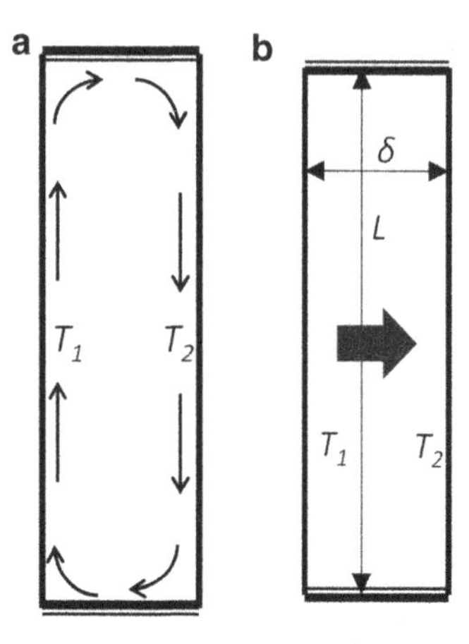

Fig. 6.7 Enclosed space with a hot left surface and a cool right surface. Upper and lower surfaces are here assumed adiabatic. (**a**) Flow pattern (**b**) Nomenclature

The characteristic length L in Table 6.3 denotes the vertical height of a *vertical enclosure*, see Fig. 6.7b. Note that the influence of convection heat transfer is negligible when enclosure is small, $Gr_\delta < 2000$. The air conductivity vs. the mean temperature according to Eq. 6.1 is used when deriving the expressions in the right column of Table 6.4. These values are then used in Eq. 6.35 to calculate the heat flux by convection across enclosed spaces.

Example 6.7 Calculate the apparent conductivity and heat transfer by convection between two parallel vertical surfaces as in Fig. 5.4 when the surfaces have temperatures $T_1 = 300\ °C = 573$ K and $T_2 = 100\ °C = 373$ K and the distance $\delta = 0.1$ m. The height $L = 0.6$ m.

Solution Equation 6.40 yields $T_\delta = \frac{(573+373)}{2} = 473$ K and Eq. 6.39 yields for air $Gr_\delta^{air} = \frac{8.11 \cdot 10^{18}(573-373) \cdot 0.1^3}{473^{13/3}} = 4.17 \cdot 10^6$. Then according to the 7th row of $Nu_\delta = 0.065 \cdot \left(4.17 \cdot 10^6\right)^{\frac{1}{3}} \cdot \left(\frac{0.6}{0.1}\right)^{-\frac{1}{9}} = 8.53$, and by inserting Eq. 6.1 into Eq. 6.37 the effective conductivity is obtained as from Eq. 6.37 $k_e = 8.53 \cdot 291 \cdot 10^{-6} \cdot 473^{0.79} \frac{\text{W}}{\text{m K}} = 0.322\ \text{W}/(\text{m K})$. The heat transfer may then be obtained from Eq. 6.36 as $\dot{q}''_c = \frac{0.322}{0.1}(573 - 373)\ \text{W}/\text{m}^2 = 644\ \text{W}/\text{m}^2$. Alternatively the explicit expression in the 4th column of the 7th row may be used.

Example 6.8 The same as Example 6.8 but in a horizontal configuration with a hot lower surface with a temperature $T_1 = 300\ °C = 573$ K and a cooler upper surface with a temperature $T_2 = 100\ °C = 373$ K, and a distance between the parallel surface $\delta = 0.1$ m, see Fig. 6.8.

Solution Equation 6.40 yields $T_\delta = \frac{(573+373)}{2} = 473$ K and Eq. 6.39 yields for air $Gr_\delta^{air} = \frac{8.11 \cdot 10^{18}(573-373) \cdot 0.1^3}{473^{13/3}} = 4.22 \cdot 10^6$. Then according to the 4th row of $Nu_\delta = 0.068 \cdot \left(4.22 \cdot 10^6\right)^{1/3} = 10.44$, and the effective conductivity of the enclosed air is obtained from Eq. 6.37 as $k_e = 10.44 \cdot 291 \cdot 10^{-6} \cdot 473^{0.79} = 0.394\ \text{W}/(\text{m K})$.

Table 6.4 Nusselt number $\boldsymbol{Nu_\delta}$ for calculating convection heat transfer in air in enclosed spaces according to Eq. 6.35 [15]

No	Orientation	Range	Nu_δ	$k_e/\delta = Nu_\delta^{air}\frac{k_f^{air}}{\delta}$
1	Horizontal layers (hotter upper layer)	(Stable layers)	1	$\frac{291 \cdot 10^{-6}\, T_\delta^{0.79}}{\delta}$
2	Horizontal layers (cooler upper layer)	$Gr_\delta < 10^4$	1	$\frac{291 \cdot 10^{-6}\, T_\delta^{0.79}}{\delta}$
3		$< 400 \cdot 10^3$	$0.195\, Gr_\delta^{1/4}$	$\frac{1.42\, (T_1 - T_2)^{0.25}}{T_\delta^{0.09} \cdot \delta^{0.25}}$
4		$Gr_\delta > 400 \cdot 10^3$	$0.068\, Gr_\delta^{1/3}$	$\frac{39.2\, (T_1 - T_2)^{0.33}}{T_\delta^{0.65}}$
5	Vertical layers	$Gr_\delta < 20 \cdot 10^3$	1	$\frac{291 \cdot 10^{-6}\, T_\delta^{0.79}}{\delta}$
6		$20 \cdot 10^3 < Gr_\delta < 200 \cdot 10^3$	$0.18\, Gr_\delta^{1/4}\left(\frac{L}{\delta}\right)^{-1/9}$	$\frac{2.80\, (T_1 - T_2)^{0.25}}{T_\delta^{0.09} \cdot L^{0.11} \cdot \delta^{0.14}}$
7		$200 \cdot 10^3 < Gr_\delta < 11 \cdot 10^6$	$0.065\, Gr_\delta^{1/3}\left(\frac{L}{\delta}\right)^{-1/9}$	$\frac{37.5\, (T_1 - T_2)^{0.33} \cdot \delta^{0.11}}{T_\delta^{0.65} \cdot L^{0.11}}$

The last column shows explicitly the thermal resistance k_e/δ

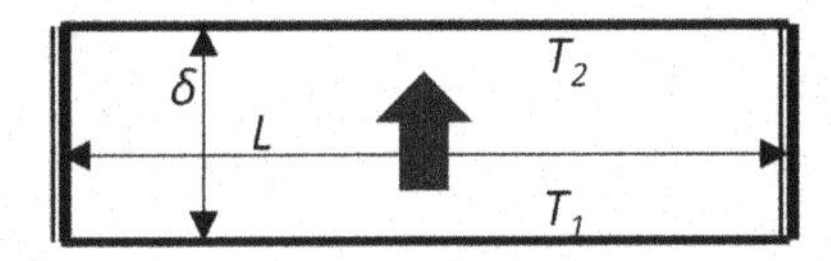

Fig. 6.8 Nomenclature for calculating heat transfer by convection across an enclosed space with two horizontal parallel surfaces of different temperatures where $T_1 > T_2$

The heat transfer may then be obtained from Eq. 6.36 as $\dot{q}''_c = \frac{0.116}{0.1}(573 - 373)\ \mathrm{W/m^2} = 232\ \mathrm{W/m^2}$. Alternatively the 4th column may be used.

Chapter 7
Numerical Methods

The analytical methods outlined in Chaps. 2 and 3 presume that the material properties and heat transfer coefficients are constant. That is, however, not possible in most cases in fire protection engineering as the temperature then varies within a wide range and therefore both material properties and boundary conditions vary considerably. Phase changes or latent heat due to water vaporization or chemical reactions of materials (see Sect. 14.1 on concrete) must in many cases be considered to achieve adequate results. Furthermore in particular radiation heat transfer coefficients vary considerably with temperature. As shown in Sect. 4.1 it increases with the third power of the temperature level. In addition geometries being considered are not as simple as assumed above. Often they are in two or three dimensions, and then analytical methods can seldom be used for practical temperature analyses. Therefore numerical methods involving computer codes are frequently used in fire protection engineering. In some cases in particular for 0-dimension problems (lumped-heat-capacity) relatively simple so-called spreadsheet codes such as Excel may be used. For problems with more complex geometries and boundary conditions computer codes based on finite difference or finite elements methods are needed. Several computer codes based on these methods are commercially available, see Sect. 7.3.2. The superposition technique as presented in Sect. 7.2 may be seen as a combination of a numerical and an analytical method.

7.1 Lumped-Heat-Capacity

The basic theory of heat transfer to bodies with uniform temperature is given in Sect. 3.1. According to Eq. 3.2

U. Wickström, *Temperature Calculation in Fire Safety Engineering*,
DOI 10.1007/978-3-319-30172-3_7

$$\frac{dT}{dt} = \frac{A}{V \cdot \rho \cdot c} \dot{q}'' \tag{7.1}$$

where A is exposed area, V volume, ρ density and c specific heat capacity of the exposed body. By integrating over time the body temperature becomes

$$T = T_i + \frac{A}{V \cdot c \cdot \rho} \int_0^t \dot{q}'' dt \tag{7.2}$$

where T_i is the initial temperature. In case $\dot{q}''$ is given as function of time (2nd kind of BC) or proportional to the difference between the surrounding and the body (surface) temperatures (3rd kind of BC), the temperature T can sometimes be solved analytically. In most other cases numerical methods must be used even when lumped heat is assumed.

In general both space and time are discretized except for lumped-heat-capacity problems with only one unknown temperature where only time is discretized.

The time derivative of Eq. 7.1 is approximated by the differential, i.e. $\frac{dT}{dt} \approx \frac{\Delta T}{\Delta t}$. Given the time is divided into increments as indicated in Fig. 7.1

$$\Delta t^{j+1} = t^{j+1} - t^j \tag{7.3}$$

and temperature increments are defined as

$$\Delta T^{j+1} = T^{j+1} - T^j \tag{7.4}$$

Assuming the time increment constant Eq. 7.1 can be written as a finite difference equation as

$$T^{j+1} - T^j = \Delta T^{j+1} \approx \frac{A}{V(c \cdot \rho)} \dot{q}''^{j}_{tot} \Delta t \tag{7.5}$$

where $\dot{q}''^{j}_{tot}$ is the heat flux to the surface at the time increment j.

In the simplest case the heat flux is proportional to the difference between the insulation surface temperature and uniform body temperature as shown in Fig. 3.3b. Then Eq. 7.5 can be written as

$$T^{j+1} = T^j + \frac{A}{V(c\rho)} \cdot \frac{k_{in}}{d_{in}} \left(T_s - T^j\right) \Delta t \tag{7.6}$$

where k_{in} and d_{in} are the conductivity and thickness of the insulation. When solved according to this forward difference scheme, all the parameters may be updated at each time step depending on the temperature of time increment j.

When a body is exposed to a third kind of boundary condition, i.e. a function of incident radiation (or radiation temperature) and gas temperature *and* the current

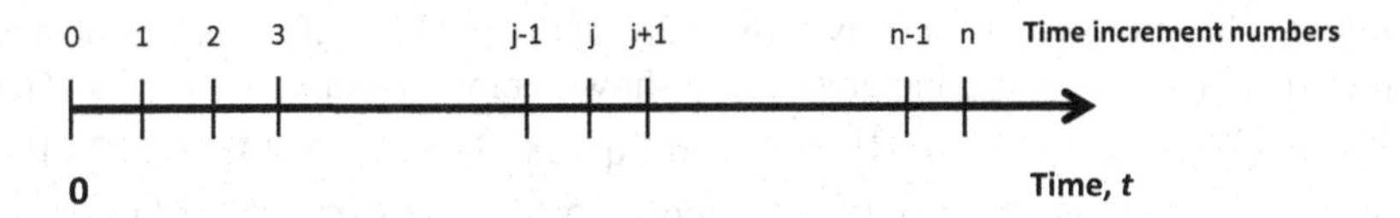

Fig. 7.1 Time axis indicating time increment numbering

surface temperature, the total heat flux $\dot{q}''_{tot}$ is defined according to Eq. 4.11 and the body temperature may be calculated as

$$T^{j+1} = T^j + \frac{A}{V(c\rho)}\left[\varepsilon\left(\dot{q}''^{j}_{inc} - \sigma\, T^{j\,4}\right) + h_c\left(T_g^j - T^j\right)\right]\Delta t \tag{7.7}$$

or when the heat flux is defined according to Eq. 4.12 as

$$T^{j+1} = T^j + \frac{A}{V(c\rho)}\left[\varepsilon\,\sigma\left(T_r^{j\,4} - T^{j\,4}\right) + h_c\left(T_g^j - T^j\right)\right]\Delta t \tag{7.8}$$

The recursion formulas of Eq. 7.7 through Eq. 7.10 are forward difference or explicit schemes. That means all parameters on the right-hand side of the equation are known at time increment j and the new temperature at time j + 1 can be calculated explicitly. Such integration schemes are numerically stable only if each time increment is chosen less than a critical time increment, the critical time increment Δt_{cr} defined as

$$\Delta t_{cr} = \frac{V \cdot c \cdot \rho \cdot d_{in}}{k_{in} \cdot A} \tag{7.9}$$

for Eq. 7.6, and

$$\Delta t_{cr} = \frac{V \cdot c \cdot \rho}{h_{tot} \cdot A} \tag{7.10}$$

for Eq. 7.7 and Eq. 7.10. h_{tot} is the total adiabatic heat transfer coefficient as defined by Eq. 4.19. The critical time corresponds to the time constant as defined in Sect. 3.1. It can vary over time as the including parameter changes with temperature. In reality much shorter time increments in the order of 10 % and the critical time increment are in general recommended to achieve accurate temperatures.

Example 7.1 An unprotected steel section with a section factor $\frac{V}{A} = 100\,\text{m}^{-1}$ is suddenly exposed to a constant fire temperature $T_f = T_r = T_g = 1000\,°\text{C}$. Calculate the steel temperature as a function of time if the initial temperature is $T_i = 20\,°\text{C}$. Assume a steel surface emissivity of 0.9 and a convection heat transfer coefficient of 25 W/(m^2 K). $\rho_{st} = 7850\,\text{kg/m}^3$ and $c_{st} = 560\,\text{Ws/(kg K)}$.

Solution This problem is ideally solved by applying Eq. 7.8 in a spreadsheet application. The first three increments are shown below. Assume $\Delta t = 10\,\text{s}$. Then $\frac{(A/V)\cdot\Delta t}{c\cdot\rho} = \frac{100\cdot 10}{560\cdot 7850} = 0.227\cdot 10^{-3}\,\text{m}^2/\text{W}$ and Eq. 7.8 yields $T^1 = 20 + 0.227\cdot 10^{-3}$ $\left[0.8\cdot 5.67\cdot 10^{-8}\left(1273^{\,4} - 298^4\right) + 25(1000-20)\right] = 52.5\,°\text{C},\ T^2 = 79.6\,\text{K}$ and $T^3 = 112\,\text{K}$. The maximum h_{tot} can be obtained from Eq. 4.19 or Fig. 4.2a as 470 W/(m^2 K) and from Eq. 7.10 the minimum increment $\Delta t_{cr} = \frac{560\cdot 7850}{100\cdot 470} = 93\,\text{s}$. At preceding lower steel temperature levels h_{tot} is much greater and thereby Δt_{cr} is much smaller, and therefore $\Delta t = 10\,\text{s}$ will yield accurate steel temperatures.

Example 7.2 A steel plate with a thickness of $d_{st} = 10\,\text{mm}$ and an initial temperature of (=293 K) is placed in the sample holder of a cone calorimeter, see Fig. 7.2. The incident radiation of the cone is set to 50 kW/m^2. The plate is well insulated on all surfaces except the upper exposed surface. Assume a steel surface emissivity of 0.9 and a convection heat transfer coefficient $h_c = 12\,\text{W}/(\text{m}^2\cdot\text{K})$. $\rho_{st} = 7850\,\text{kg}/\text{m}^3$ and $c_{st} = 560\,\text{Ws}/(\text{kg K})$. Derive a time integration scheme and show the first three time increments.

Solution Apply Eq. 7.7 where $\frac{A}{V\rho c} = \frac{1}{d\rho c} = \frac{1}{0.01\cdot 7850\cdot 560} = \frac{1}{44000}\,\text{Ws}/\text{m}^2\text{K}$ and $\dot{q}''^{\,j}_{tot} = 0.9\left(50000 - \sigma T^{4,j}\right) + 12\cdot(20 - T)$. Thus a forward difference incremental scheme becomes $T^{j+1} = T^j + \frac{\Delta t}{44000}\left\{0.9\left(50000 - \sigma T^{4,j}\right) + 12\cdot\left(293 - T^j\right)\right\} = \frac{(0.9\cdot 50000 + 12\cdot 293)\Delta t}{44000} + \left(1 - \frac{12\ \Delta t}{44000}\right)T^j - \frac{\Delta t}{44000}\cdot 5.67\cdot 10^{-8} T^{4,j}\big)$. Assume a time increment $\Delta t = 60\,\text{s}$. Then the incremental scheme can be reduced to: $T^{j+1} = 66.2 + 0.984\cdot T^j - 77.3\cdot 10^{-12}\cdot T^{4,j}$.

First step, j = 1: $T^2 = 66.2 + 0.984\cdot 293 - 77.3\cdot 10^{-12}\cdot 293^4 = 66.2 + 288 - 0.57 = 354\ \text{K} = 80.6\,°\text{C}$.

Second step, j = 2: $T^3 = 66.2 + 0.984\cdot 354 - 77.3\cdot 10^{-12}\cdot 354^4 = 66.2 + 349 - 1.2 = 413\ \text{K} = 141\,°\text{C}$.

Third step, j = 3: $T^4 = 66.2 + 0.984\cdot 413 - 77.3\cdot 10^{-12}\cdot 413^4 = 66.2 + 406 - 2.25 = 470\ \text{K} = 197\,°\text{C}$.

Comment: Figure 7.3 shows a comparison between measured and calculated temperature steel specimen as in the example. In the calculations Eq. 7.8 was applied with a heat transfer coefficient h_c increased to 18 W/(m^2 K) to consider the heat losses from the steel specimen by conduction and a gas temperature T_g as measured with a thin thermocouple. The accurate prediction indicates how well Eq. 7.8 models the heat transfer to a specimen surface in the cone calorimeter.

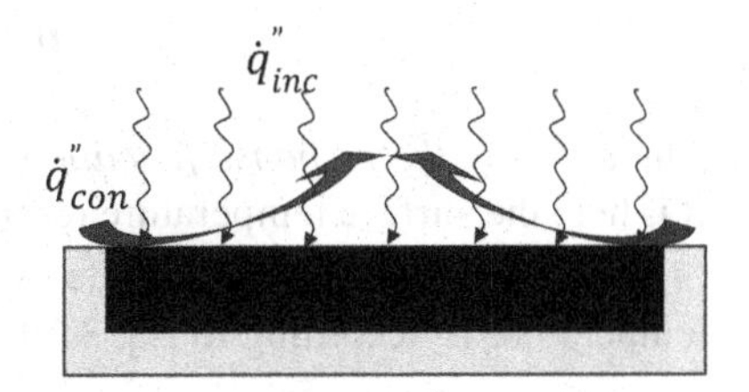

Fig. 7.2 Heat transfer to a 10-mm-thick steel plate in the cone calorimeter

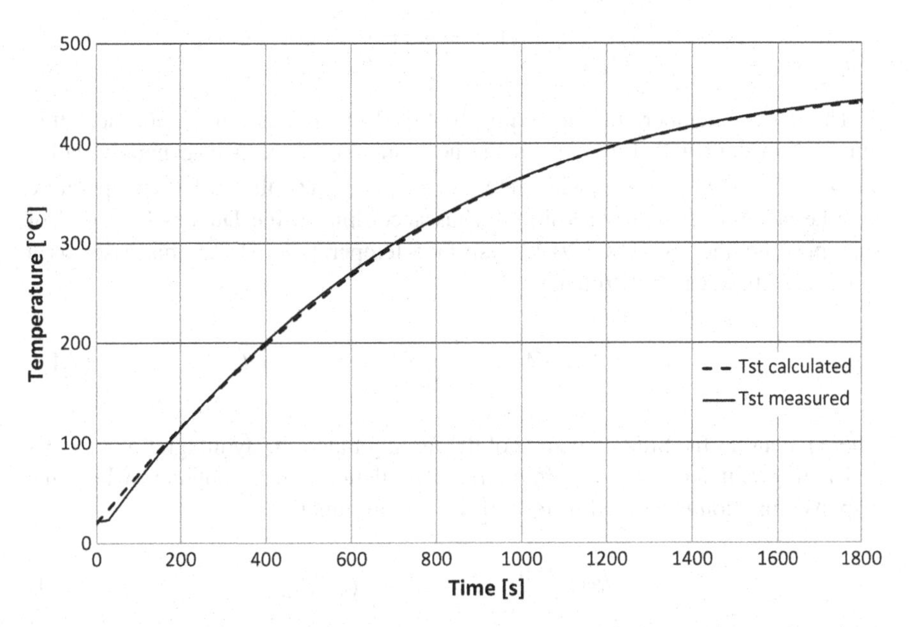

Fig. 7.3 Comparison of measured and calculated temperature of 10-mm-thick steel specimen exposed to an incident radiation of 50 kW/m^2 in a cone calorimeter

7.2 Superposition and the Duhamel's Superposition Integral

A technique based on superposition is presented below. In its infinitesimal form it may be called *Duhamel's superposition integral*. It is a technique which has many various types of applications when analysing bodies with *constant material properties* and with *zero initial temperature conditions* which are exposed to boundary conditions varying with time. Zero initial temperature conditions can be obtained for bodies with constant initial temperatures by calculating temperature rise as shown below.

The technique is here exemplified for the case of a surface of a semi-infinite solid at uniform temperature T_i. See also [1, 2]. When it is suddenly receiving a *constant* external heat flux f at the time $t = 0$, the surface temperature rise $\theta_s = T_s - T_i$ may then be written as

$$\theta_s = f \cdot A(t) \tag{7.11}$$

where *A(t)* is *the response function* (sometimes called the *fundamental solution*). It is here the surface temperature response as a function of time for a unit heat flux ($f = \dot{q}''_{tot} = 1$). For semi-infinite solids the response function of the surface temperature is according to Eq. 3.29

$$A(t) = \frac{2\sqrt{t}}{\sqrt{\pi}\sqrt{(k\rho c)}} \tag{7.12}$$

The surface temperature according to Eq. 7.11 applies only if the heat flux remains constant with time. When the heat flux to the surface, generally called the *forcing function*, varies with time, i.e. $f(t) = \dot{q}''_{tot}(t)$, the surface temperature may be calculated by superposition. Thus according to the Duhamel integral of superposition the solution $R(t)$ (the surface temperature rise in this case) as a function of time can be written as

$$R(t) = f(0) \cdot A(t) + \int_{\xi=0}^{t} f'(t - \xi) \cdot A(\xi)\, d\xi \tag{7.13}$$

where f' denotes the time derivative of the forcing function. By integration by parts and noting that $A(t) = 0$, an alternative formulation can be obtained where the response function is derived instead of the forcing function.

$$R(t) = \int_{\xi=0}^{t} f(t - \xi) \cdot A'(\xi)\, d\xi \tag{7.14}$$

where A′ is the time derivative of the response function. (ξ is a dummy variable defined only within the integral.)

Equation 3.16 in Sect. 3.1.2.1 giving the temperature of a thermocouple modelled as lumped heat was derived from Eq. 7.14. In that case the response function can be derived analytically and depending on the forcing function, the integral can in some cases be solved analytically. In other cases numerical integration techniques must be used to calculate the thermocouple temperature at a given time *t*.

Then the time is divided into increments and the surface temperature rise T_i, i.e. $\theta = T_s - T_i$, may be calculated numerically by a time step superposition scheme. To illustrate how solutions can be superimposed to obtain the surface temperature of a time-dependent flux, the following case is studied. The heat flux is assumed to vary as shown in Fig. 7.4. Thus

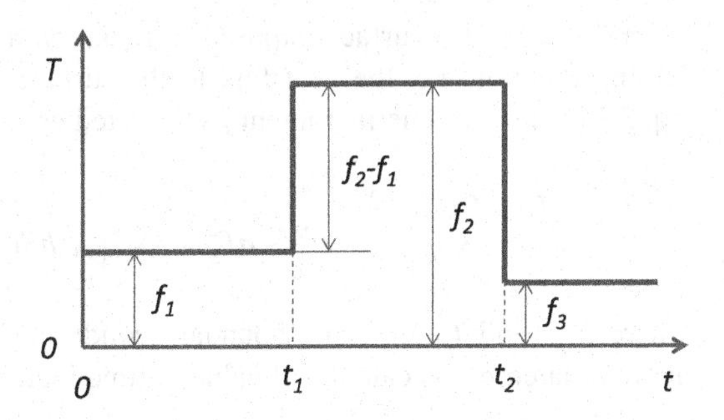

Fig. 7.4 A stepwise changing forcing function

$$\begin{aligned} f(t) &= q_1 \quad 0 < t < t_1 \\ f(t) &= q_2 \quad t_1 < t < t_2 \\ f(t) &= q_3 \quad t > t_2 \end{aligned} \tag{7.15}$$

Then the surface temperature rise can be calculated as

$$\theta_s = q_1(t) \cdot A(t) + (q_2 - q_1) \cdot A(t - t_1) + (q_3 - q_2) \cdot A(t - t_2) \tag{7.16}$$

In a general form the surface temperature rise may be written as

$$\theta_s = f(0) \cdot A(t) + \sum_{i=1}^{n} \Delta f(t_i) \cdot A(t - t_i) \tag{7.17}$$

where

$$\Delta f(t_i) = f(t_{i+1}) - f(t_i) \tag{7.18}$$

A very powerful superposition technique is shown below which allows the forcing function to depend on the response for the actual exposure. That is, for instance, the case when a surface is exposed to radiation and the emitted radiation depends on the surface temperature. Then a new surface temperature rise can then be calculated at time increment $j + 1$ as:

$$\theta_{j+1} = \sum_{i=0}^{j} \left[(f_i - f_{i-1}) \cdot A_{j-i} \right] \tag{7.19}$$

or alternatively as

$$\theta_{j+1} = \sum_{i=0}^{j} \left[f_{j-i} \cdot (A_{i+1} - A_i) \right] \tag{7.20}$$

where Δt is the time increment and $\theta_0 = f_0 = 0$. The values of the time-dependent parameters at a given number of time increments i are defined as $\theta_i = \theta(i \cdot \Delta t)$, $f_i = f(i \cdot \Delta t)$ and $A_i = A(i \cdot \Delta t)$.

In the case of a surface exposed to radiation and convective heat transfer, the forcing function, i.e. the heat flux to the surface, can be calculated according to Eq. 3.14. Then at time increment j calculated as

$$\begin{aligned} f(j \cdot \Delta t) &= \dot{q}''_{tot}(j \cdot \Delta t) \\ &= \varepsilon\left[\dot{q}''_{inc}(j \cdot \Delta t) - \sigma T_s^4(j \cdot \Delta t)\right] + h\left(T_g(j \cdot \Delta t) - T_s(j \cdot \Delta t)\right) \end{aligned} \quad (7.21)$$

where $\dot{q}''_{inc}$ and T_g are input boundary conditions and T_s the surface temperature approximated by its calculated value at time increment j.

Example 7.3
A concrete wall surface is assumed to receive piecewise constant heat fluxes (boundary condition of second kind) according to Eq. 7.15 and as indicated in Fig. 7.4 with the following input values:

$t_1 = 10\,\text{min}$, $t_2 = 20\,\text{min}$ and $q_1 = 20\,\text{kW}/(\text{m}^2\text{K})$, $q_2 = 35\,\text{kW}/(\text{m}^2\text{K})$ and $q_3 = 15\,\text{kW}/(\text{m}^2\text{K})$

The initial wall temperature is 20 °C. Assume thermal properties of concrete according to Table 1.2.

Express the surface temperature rise as a function of time for the three time intervals by superposition according to Eq. 7.16.

Solution According to Table 1.2 $(k \cdot \rho \cdot c) = 3.53 \cdot 10^6 \left(\text{W}^2\text{s}\right)/\left(\text{m}^4\text{K}^2\right)$. Then Eq. 7.16 yields with the response function according to Eq. 3.29 new surface temperature rises as:

Time interval [s]	Surface temperature rise, θ_s [°C]
$0 < t < 600$	$20000 \cdot \frac{2\sqrt{t}}{\sqrt{\pi}\sqrt{3.53 \cdot 10^6}}$
$600 < t < 1200$	$20 \cdot 10^3 \cdot \frac{2\sqrt{t}}{\sqrt{\pi}\sqrt{3.53 \cdot 10^6}} + (35-20) \cdot 10^3 \cdot \frac{2\sqrt{t-600}}{\sqrt{\pi}\sqrt{3.53 \cdot 10^6}}$
$t > 1200$	$20 \cdot 10^3 \cdot \frac{2\sqrt{t}}{\sqrt{\pi}\sqrt{3.53 \cdot 10^6}} + (35-20) \cdot 10^3 \cdot \frac{2\sqrt{t-600}}{\sqrt{\pi}\sqrt{3.53 \cdot 10^6}} + (15-35) \cdot 10^3 \cdot \frac{2\sqrt{t-1200}}{\sqrt{\pi}\sqrt{3.53 \cdot 10^6}}$

7.3 The Finite Element Method for Temperature Analyses

When calculating temperature in fire-exposed structures, non-linearities must in most cases be considered. The boundary conditions are non-linear varying significantly with temperature as shown above (see Chap. 4), and also the thermal properties of most materials vary significantly within the wide temperature span that must be considered in FSE problems. Therefore numerical methods must usually be employed. The most general and powerful codes are based on the so-called finite element method (FEM). Below the basic equations are derived for

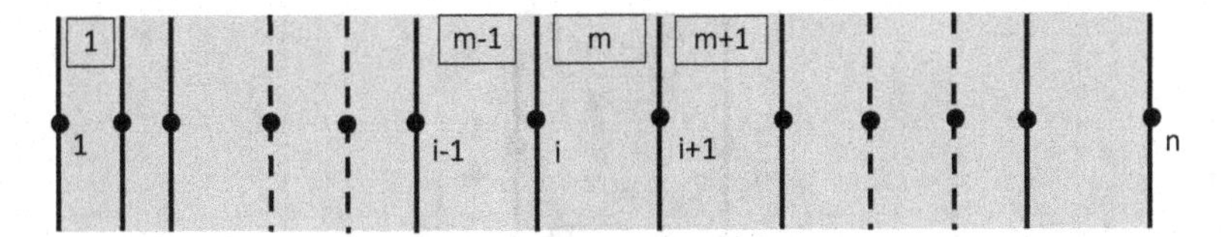

Fig. 7.5 A wall divided into one-dimensional elements numbered with *m:s* and with the nodes numbered with *i:s*

a simple one-dimensional case as an illustration. Similar types of equation may be derived for two and three dimensions.

7.3.1 One-Dimensional Theory

Figure 7.5 shows a wall which has been divided into a number of one-dimensional elements. The temperature between the nodes is assumed to vary linearly along the length.

In any element, interior or at the surface, with the length L, see Fig. 7.6, the conductivity k and a cross-section area A, the heat flow to the element nodes can be calculated as

$$\dot{q}_1 = \frac{k}{L} \cdot (T_1 - T_2) \tag{7.22}$$

and

$$\dot{q}_2 = -\frac{k}{L} \cdot (T_1 - T_2) \tag{7.23}$$

or in matrix format as

$$\bar{\dot{q}}^e = \bar{\dot{k}}^e \cdot \bar{\dot{T}}^e \tag{7.24}$$

where $\bar{\dot{q}}^e$ is the element node heat flow vector, $\bar{\dot{k}}^e$ the element thermal conduction matrix and $\bar{\dot{T}}^e$ the element node temperature vector. Given the one-dimension assumption, the cross-section area is constant and assumed equal unity. Then the element thermal conduction matrix may then be identified as

$$\bar{k}^e = \begin{bmatrix} k_{11}^e & k_{12}^e \\ k_{21}^e & k_{22}^e \end{bmatrix} = \frac{k}{L} \begin{bmatrix} +1 & -1 \\ -1 & +1 \end{bmatrix} \tag{7.25}$$

and the element nodal temperature and heat flow vectors as

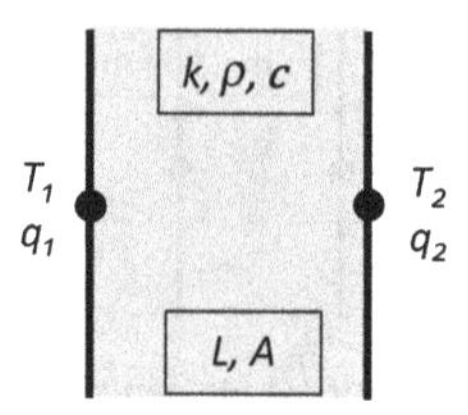

Fig. 7.6 A one-dimensional element with local element node numbers 1 and 2 and with a length L and a section area A. The element is given a thermal conductivity k, a specific heat capacity c and a density ρ

$$\dot{\overline{T}}^e = \begin{Bmatrix} T_1 \\ T_2 \end{Bmatrix} \tag{7.26}$$

and

$$\dot{\overline{q}}^e = \begin{Bmatrix} q_1 \\ q_2 \end{Bmatrix} \tag{7.27}$$

In a similar way an element heat capacity matrix can be defined by lumping the heat capacity of the element in the nodes. Thus an element heat capacity matrix may be obtained as

$$\overline{c}^e = \frac{L \cdot c \cdot \rho}{2} \begin{bmatrix} 1 & 0 \\ 0 & 1 \end{bmatrix} \tag{7.28}$$

When several elements are combined, global heat conductivity matrix $\overline{K}$ can be assembled. In the very simple case of three one-dimensional elements, the global thermal conduction matrix becomes

$$\overline{K} = \begin{bmatrix} k_{11}^1 & k_{12}^1 & 0 & 0 \\ k_{21}^1 & \left(k_{22}^1 + k_{11}^2\right) & k_{12}^2 & 0 \\ 0 & k_{21}^2 & \left(k_{22}^2 + k_{11}^3\right) & k_{12}^3 \\ 0 & 0 & k_{21}^3 & k_{22}^3 \end{bmatrix} \tag{7.29}$$

where the super fixes 1–3 denote the contributions from the corresponding element numbers. The global heat capacity matrix $\overline{C}$ may be assembled in a similar way as the global conductivity matrix. Notice that both the heat conductivity and the heat capacity matrices are symmetric and dominated by their diagonal elements, and that the global heat capacity matrix assembled from element matrices according to Eq. 7.28 will have non-zero elements only in the diagonal. This will have a decisive influence on how global algebraic heat balance equation can be solved as shown below.

In global form the heat balance equation may now be written in matrix form as

$$\overline{C}\dot{\overline{T}} + \overline{K}\,\overline{T} = \overline{Q} \tag{7.30}$$

where the vector $\dot{\overline{T}}$ contains the time derivatives of the node temperatures. Each row in this equation system represents the heat balance of a node. For each equation or each node either the temperature or the heat flow given in the corresponding rows in the vectors $\overline{T}$ and $\overline{Q}$, respectively, are known. In principle three cases are possible for each equation/row (c.f. the three kinds of boundary conditions as presented in Chap. 4):

1. The node temperature T_i is prescribed.
2. The node heat flow Q_i is prescribed.
3. The node heat flow Q_i can be calculated as a function of a given gas temperature and radiation temperature, and the surface temperature.

In the first case the corresponding equation vanishes as the unknown quantity is prescribed a priori. The most common case for internal nodes is the second case, i.e. the external flow is zero. A typical boundary condition when calculating temperature in fire-exposed structures is of the third case corresponding to a boundary of the 3rd kind. Then according to Table 4.1 of Chap. 4 the nodal heat flow is

$$\dot{Q}_i = \varepsilon\sigma\left(T_r^4 - T_{s,i}^4\right) + h\left(T_g - T_{s,i}\right) \tag{7.31}$$

(given the cross-section area equal unity). Notice that this is non-linear as the emitted radiation depends on the temperature raised to the fourth power. This is of importance when choosing the equation-solving methodology.

The differential global matrix equation Eq. 7.30 is solved numerically by approximating the time derivative of the node temperatures as

$$\dot{\overline{T}} \approx \frac{\Delta\overline{T}}{\Delta t} = \frac{\overline{T}^{j+1} - \overline{T}^{j}}{\Delta t} \tag{7.32}$$

where $\overline{T}^j$ is the node temperature vector at time step j and Δt is here a chosen time increment. Now the heat balance equation in matrix format Eq. 7.30 can be written as

$$\overline{C}\left[\frac{\overline{T}^{j+1} - \overline{T}^{j}}{\Delta t}\right] + \overline{KT} = \overline{Q} \tag{7.33}$$

In this differential equation the temperature vector is known at time increment j. The new temperature vector at time $j + 1$ is obtained explicitly based on the conditions at time step j for calculating the thermal conduction as

$$\overline{T}^{j+1} = T^j + \Delta t\, \overline{C}^{-1}\left[\overline{Q}^j - \overline{K}\overline{T}^j\right] \tag{7.34}$$

As the heat capacity matrix is here assumed diagonal (c.f. Eq. 7.28), the new node temperatures at time step $j+1$ can be obtained directly row by row and no system equation needs to be solved. Alternatively an implicit method may be derived where the conduction is based on the temperatures at time step $j+1$. Then the new node temperatures may be calculated as

$$\overline{T}^{j+1} = \left(\frac{\overline{C}}{\Delta t} + \overline{K}\right)^{-1} \left(\overline{Q}^j + \overline{C}\overline{T}^j\right)\Delta t \tag{7.35}$$

Combinations of the two solution methods are also possible but as soon as the conduction depends on the node temperature at time step $j+1$ the solution scheme requires the solution of a global equation system containing as many unknowns as there are unknown node temperatures. Most finite element computer codes use this type of implicit solution schemes. They are generally numerically more stable than the explicit techniques and therefore longer time increments may be used.

The explicit solution according to Eq. 7.34 may, on the other hand, be very simple when the heat capacity matrix $\overline{C}$ is diagonal, i.e. it contains only non-zero elements in the diagonal as shown for a one-dimensional element in Eq. 7.28. The solution of the equation system becomes then trivial as each nodal temperature can be obtained directly/explicitly, one at a time. It involves only a multiplication of a matrix with a vector which requires much less computational efforts than solving an equation system. This solution scheme is, however, numerically stable only when the time increment Δt is less than a critical value proportional to the specific heat capacity times the density over the heat conductivity of the material times the square of a characteristic element length dimension Δx. This requirement applies to all the equations of the entire system, all nodes i except those with prescribed temperatures. If violated in any of the equations, i.e. at any point of the finite element model, the incremental solution equation will become unstable (cf. Sect. 7.1 and Eq. 7.10 on lumped-heat-capacity). Hence in the one-dimensional case treated here the *critical time increment* Δt_{cr} may be estimated as

$$\Delta t_{cr} \approx \min\left[\frac{\rho \cdot c}{2k}\,\Delta x^2\right]_i \tag{7.36}$$

This means that short time increments are needed for materials with a low density and a high conductivity, and when small element sizes are used. At boundary nodes with heat transfer conditions of the 3rd kind the critical time increment will be influenced by the heat transfer coefficient h as well. Then at any node i

$$\Delta t_{cr} \approx \min \left[\frac{\rho \cdot c}{k + h/\Delta x} \Delta x^2 \right]_i \tag{7.37}$$

The heat transfer coefficient h is here the sum of the heat transfer coefficient by convection and radiation, denoted by h_{tot} in Sect. 4.3.

In practice, when calculating temperature in fire-exposed structures, short time increments must be used independent of solution technique as the duration of analyses are short and boundary condition chances fast. Therefore numerical stability is only a problem when modelling sections of very thin metals sheets with high heat conductivity. Then very short time increments are required. The problem may, however, be avoided by prescribing that nodes close to each other shall have the same temperature. This technique has been applied in the code TASEF [14]. In that code a technique is also developed where the critical time increment is estimated and thereby acceptable time increments can be calculated automatically at each time step depending on thermal material properties and boundary conditions varying with temperature. At boundaries of the 3rd kind short time critical time increments can be avoided by assuming the surface temperature equal to the surrounding temperature (boundary of the 2nd kind). This approximation may be applied when the thermal inertia of a material is relatively low and the surface temperature is expected to follow close to the exposure temperature (adiabatic temperature).

As a general rule finite element calculations shall not be accepted until it is shown that the solution gradually converge when time increments and element sizes are reduced. This rule applies to both computer codes using explicit and implicit solution techniques. A guidance standard on requirements for calculation methods that provide time-dependent temperature field information resulting from fire exposures required for engineered structural fire design has been published by SFPE [18].

7.3.2 *Computer Codes for Temperature Calculations*

Several computer codes are commercially available for calculating temperature in fire-exposed structures. They are in general based on the finite element method. Some are specifically developed and optimized for calculating temperature in fire-exposed structures while others are more general purpose codes.

TASEF [14, 19] and SAFIR [20] are examples of programs which have been developed for fire safety problems. They have different pros and cons. They all allow for temperature-dependent material properties and boundary conditions. TASEF employs a forward difference solving technique which makes it particularly suitable for problems where latent heat due to, e.g. vaporization of water must be considered. It yields also in most cases very short computing times, in particular for problems with a large number of nodes. TASEF and SAFIR have also provisions for modelling heat transfer by convection and radiation in internal voids. TASEF does also allow for boundary conditions where the exposure radiation and gas temperature are different, boundary condition of the 3rd kind according to Table 4.1.

There are many very advanced general purpose finite element computer codes commercially available such as ABAQUS, ANSYS, ADINA, Heating 7 and Comsol. The main advantage of using this type of codes is that they have several types of elements for various geometries and dimensions, and that they come with advanced graphical user interfaces and pre- and post-processors.

7.3.3 On Accuracy of Finite Element Computer Codes

There are at least three steps that must be considered when estimating the accuracy of computer codes for numerical temperature calculations:

1. Accuracy of material properties
2. Verification of the calculation model
3. Validity of the calculation model

The first point is crucial. Errors in material property input will be transmitted into output uncertainties and errors. Methods for measuring material properties at high temperature are briefly discussed in Sect. 1.3.3.

Secondly, the numerical verification of the computer code itself is important. *Verification* is the process of determining the degree of accuracy of the solution of the governing equations. Verification does not imply that the governing equations are appropriate for the given fire scenario, only that the equations are being solved correctly.

The third point is of course important as well. *Validation* is the process of determining the degree to which a mathematical model and a calculation method adequately describe the physical phenomena of interest. Temperature calculation codes are in general developed for solving the Fourier heat transfer equation. Effects of varying material thermal properties can be considered in the numerical integration while, for example, the thermal effects of spalling or water migration cannot generally be predicted. Other important aspects are the possibilities of satisfactory describing boundary conditions. For FSE problems generally involving high temperatures, the calculation of heat transfer by radiation at external boundaries and in internal voids is of special concern.

The codes mentioned above yield results with acceptable accuracy for simple well-defined boundary conditions and material properties. Differences when mixed boundary conditions and latent heat are introduced. A scheme to follow including a number of reference cases of various levels of complexity has been published in an SFPE standard [21]. Precisely calculated reference temperatures of 16 cases of bodies have been listed. They represent a variety of problems that are relevant in FSE involving a range of complexities.

Some reference cases are linear problems which can be solved analytically. Then when increasing the number of elements the results should converge to one correct value. Codes yielding results that converge smoothly when increasing the number of elements are generally more reliable for the type of problems considered. Most of the reference cases are relevant for FSE including effects of conductivity varying with temperature, latent heat, radiant heat transfer boundary conditions and combinations of materials, concrete, steel and mineral wool. Then as no exact analytical solutions are available, the cases were modelled in the finite element codes Abaqus and TASEF. The difference between the solutions obtained with these codes were within one-tenth of a degree Celsius, and as these codes employ different calculations the published solutions of the reference cases were deemed very accurate.

7.3.4 On Specific Volumetric Enthalpy

As shown in Eq. 7.30 the heat conduction equation can be expressed in terms of *specific volumetric enthalpy e*. This is advantageous when calculating temperature with numerical methods in cases with materials where latent heat needs to be considered. The specific volumetric enthalpy or here often just the enthalpy is the heat content of a material due to temperatures above zero per unit volume (Ws/m^3), i.e.

$$e(T) = \int_0^T \rho \cdot c\, dT + \sum\nolimits_i l_i \tag{7.38}$$

where ρ is density and c specific heat capacity. These are in general temperature dependent. The second term $\sum_i l_i\,(Ws/m^3)$ represents *latent heats* required for various chemical and physical phase changes at various temperature levels. The first term is the *sensitive heat*. The most common form of latent heat to be considered in FSE is the vaporization of moisture (free water) when the temperature rise passes the boiling point (100 °C).

For a dry inert material with a density ρ_{dry} and a specific heat c_{dry} not varying with temperature the enthalpy is proportional to the temperature and the *sensitive heat* becomes

$$e = c_{dry}\rho_{dry}T \tag{7.30}$$

If a material contains free water, the enthalpy versus temperature is influenced in two ways. Firstly heat proportional to the temperature rise (sensitive heat) is needed to increase the temperature of the water, and then in addition heat (latent heat) is needed for vaporizing water at temperatures in an interval above 100 °C. Both these components must be added to the enthalpy of the dry material when calculating the enthalpy as function of temperature. Thus in general terms the enthalpy consists of three components, the sensitive heat of the dry material, the sensitive heat of the water and the latent heat due to vaporization of water. The first term is present over the entire temperature range, the second added only as long as water is present, and the third only in the temperature interval when the water vaporizes.

Moisture content u is usually expressed as the percentage water by mass of the dry material. Thus u is defined as

$$u = 100 \cdot \frac{\rho_{ori} - \rho_{dry}}{\rho_{dry}} \tag{7.40}$$

where ρ_{ori} is the original density of the moist material. If the moisture is assumed to evaporate between a lower temperature T_l and an upper T_u temperature, the latent heat due to water vaporization l_w is added to the enthalpy at the upper temperature level. The latent heat of water is then calculated as

$$l_w = \frac{u}{100} \rho_{dry} \, a_w \tag{7.41}$$

where the heat of vaporization of water $a_w = 2.26 MJ/kg$. As an example the enthalpy as a function temperature of a material with constant dry properties can then be obtained as shown in Table 7.1 and Fig. 7.7. The enthalpy is then calculated at four temperature levels and in-between the enthalpy varies linearly. Notice that as an average only half of the water is assumed to be heated between the lower temperature T_l (100 °C) and the upper temperature T_u for the vaporization process.

As an example the enthalpy of a concrete with a dry density of 2400 kg/m^3, a specific heat of 800 J/(kg K) and a moisture content $u = 3$ % by mass is shown in Fig. 7.7a. For comparison the enthalpy for a dry concrete ($u = 0$ %) is given as well. The moisture is assumed to evaporate linearly with temperature between 100 and 120 °C. Notice that at temperatures above T_u, the enthalpy rises linearly with temperature at the same rate as for a dry material.

Most computer programs require input of the specific heat and the density or the product of the two. This parameter is obtained by deriving the temperature–enthalpy curve. For the case above the specific volumetric heat c·ρ then becomes as shown in Fig. 7.7b.

The *volumetric specific heat* (c·ρ) as a function of temperature then increases suddenly in the range where the water is assumed to evaporate. This may cause numerical problems in particular for cases where the temperature range is narrow and the moisture content is high.

Gypsum is often used to seal penetrations through fire barriers and to protect steel structures. To raise the temperature of gypsum heat is needed to heat the dry material and to heat and evaporate the free water. In addition heat is needed for dehydration and vaporization of the crystalline bound water which occurs in two steps. An example of calculated specific volumetric enthalpy of gypsum containing 5 % free water and 21 % crystalline bound water is shown in Fig. 7.8 based on work

Table 7.1 Calculation of specific volumetric enthalpy, e, for a material with constant dry properties with a moisture content of u % by mass of the dry material

Temperature, T	Specific volumetric enthalpy, e
0	$e_0 = 0$
T_l	$e_{T_l} = \rho_{dry}\left[c_{dry} + \frac{u}{100}c_w\right]T_l$
T_u	$e_{T_u} = e_{T_l} + \rho_{dry}\left[c_{dry} + 0.5\frac{u}{100}c_w\right](T_u - T_l) + \frac{u}{100}\rho_{dry}\,a_w$
$T > T_u$	$e_{T>T_u} = e_{T_u} + c_{dry}\rho_{dry}(T - T_u)$

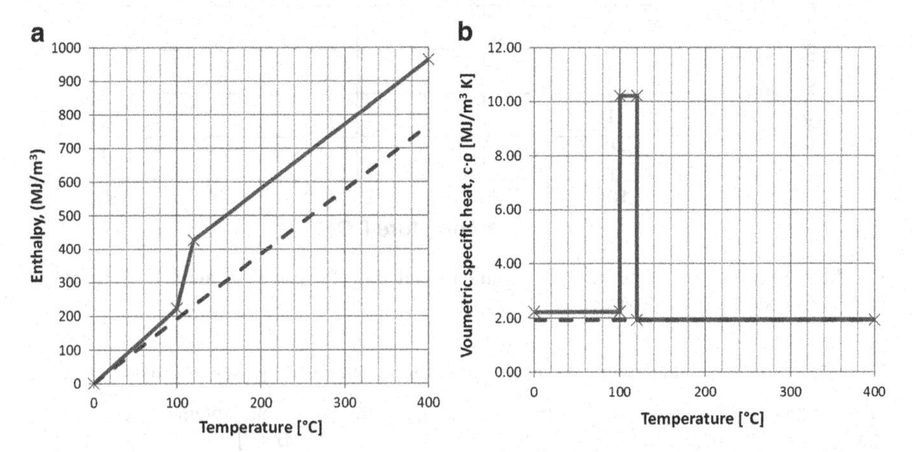

Fig. 7.7 Example of specific volumetric enthalpy vs. temperature of dry and moist concrete (3 % by mass). The moisture is assumed to evaporate linearly between 100 and 120 °C. (**a**) Specific volumetric enthalpy, e. (**b**) Specific volumetric heat capacity, $c \cdot \rho$

by Thomas [22] (the figure is taken from a master thesis of Emil Ringh (2014), Luleå TU).

Notice how the latent heats for the dehydration and vaporization processes surpass by far the sensitive heat needed to heat the inert material by comparing the slope of the curve below 100 °C thereafter until all the water has evaporated at temperatures above 220 °C. This ability of gypsum to absorb has a significant effect, for instance, on gypsum boards for fire insulation of steel structures. However, this is only for gypsum board qualities able to resist fire exposures. To take advantage of the effects in calculations, computer codes where the specific volumetric enthalpy can be input directly are the most suitable as it may be difficult to convert the curve into density and specific heat which corresponds to the derivative or slope of the curve.

Reliable values of the conductivity of gypsum are hard to find in the literature as the temperature development in gypsum depends very much on the highly non-linear enthalpy curve due to latent heats. With great reservation on the accuracy the values in Table 7.2 are recommended to be used in combination with the temperature–enthalpy curve shown in Fig. 7.8 for indicative calculations.

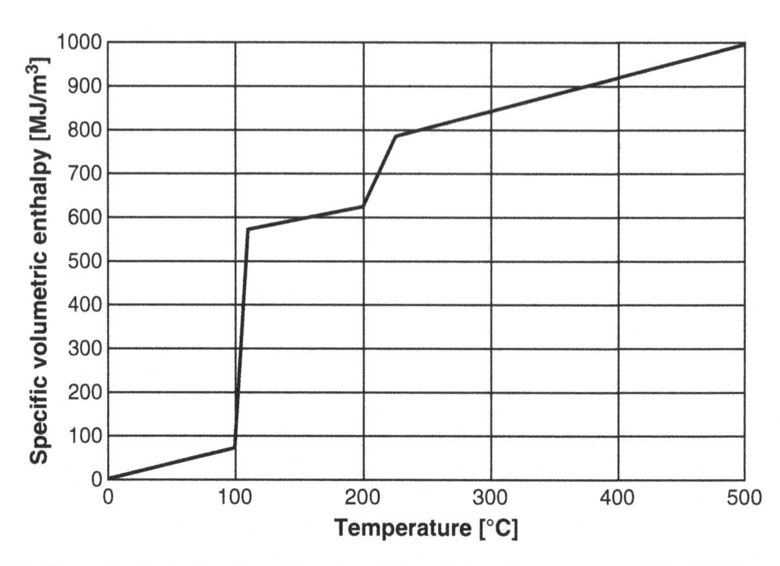

Fig. 7.8 Example of calculated specific volumetric enthalpy of gypsum containing 5 % free water and 21 % crystalline bound water

Table 7.2 Thermal conductivity and specific volumetric enthalpy of gypsum as given in Fig. 7.8

Temperature [°C]	Thermal conductivity [W/(m K)]	Temperature [°C]	Specific volumetric enthalpy [MJ/m^3]
20	0.19	0	0
100	0.15	100	73.5
500	0.17	110	571
1000	0.35	200	625
2000	0.35	220	784
		2000	2123

Chapter 8
Thermal Ignition Theory

The various aspects of the subject ignition of unwanted fires has been thoroughly investigated by Babrauskas and presented in the comprehensive *Ignition Handbook* [23]. This book is concentrating on the calculation of the development of surface temperature. Despite many limitations, it is often assumed that a solid ignites due to external heating when its exposed surface reaches a particular *ignition temperature*.

In Sect. 8.1 some data of ignition temperature of various substances are given and then in Sect. 8.2 handy formulas are presented on how to calculate time to ignition of surfaces exposed to constant incident radiation heat flux. These formula yields very similar results in comparison to accurate and elaborate numerical calculations.

8.1 Ignition Temperatures of Common Solids

Combustible solids may ignite due to *piloted ignition*, or *auto-ignition* (also called *spontaneous ignition*). The piloted ignition temperature of an externally heated substance is the surface temperature at which it will ignite in a normal atmosphere with an external source of ignition, such as a small flame or spark, present. Most common materials then ignite in the range of 250–450 °C. The auto-ignition temperature is the corresponding temperature at which a substance will spontaneously ignite without a flame or spark present. It is considerably higher, normally exceeding 500 °C.

Some limited amount of relevant material data are given in Table 8.1 for some liquids and in Table 8.2 for some plastics.

Note that the times to ignition as estimated by the thermal theories outlined below are generally very crude and based on the assumption of homogeneous materials with constant material properties not varying with temperature or time. The formulas are, however, very useful for the intuitive understanding of which

U. Wickström, *Temperature Calculation in Fire Safety Engineering*,
DOI 10.1007/978-3-319-30172-3_8

Table 8.1 Critical temperatures of some liquids

Liquid	Formula	Flash point [K]	Boiling point [K]	Auto-ignition [K]
Propane	C_3H_5	169	231	723
Gasoline	Mixture	~228	~306	~644
Methanol	CH_3OH	285	337	658
Ethanol	C_2H_5OH	286	351	636
Kerosene	~$C_{14}H_{30}$	~322	~505	~533

From Quintiere [24]

Table 8.2 Ignition temperatures of some plastics grouped by category

Category of solid	Ignition temperature [°C]	
	Piloted	Auto
Thermoplastics	369 ± 73	457 ± 63
Thermosetting plastics	441 ± 100	514 ± 92
Elastomers	318 ± 42	353 ± 56
Halogenated plastics	382 ± 79	469 ± 79

From Babrauskas [23]

material and geometrical properties govern the ignition process and the ignitability characteristics.

The time to ignition of thick homogenous materials is proportional to the thermal inertia ($k{\cdot}\rho{\cdot}c$), i.e. the product of specific heat capacity, density and conductivity, see Sect. 3.2. The conductivity increases generally at the same time as the density of a material increases (see Eq. 1.36). Therefore the thermal inertia of materials varies over a large range and consequently the ignition properties. Insulating materials have low conductivities k (by definition) and low densities ρ and will therefore ignite easily if combustible. The specific heat capacity depends on the chemical composition of the material, but the values of common materials found in the literature do not vary much. An exception is wood which according to values found in the literature has a relatively high effective specific heat capacity. (This may be a way of considering the effects of its water content.)

Table 1.2 shows how the thermal inertia increases considerably with density for various combustible and non-combustible materials. Notice for instance that the thermal inertia of an efficient insulating material such as polyurethane foam is less than a hundredth of the corresponding value of solid wood. Then as an example a low density wood fibre board may have a density of 100 kg/m^3 and a conductivity of 0.04 W/(m K), while a high density wood (oak) have a density of 700 kg/m^3 and a conductivity of 0.17 W/(m K). As such boards can be assumed to have about the same specific heat capacity, it can be calculated that the thermal inertia of the high density fibre board is more than 40 times higher of that of the low density board. The low density fibre board can therefore ideally be estimated to ignite 40 times faster than the high density fibre board when exposed to the same constant heating conditions, see Eq. 8.9.

8.2 Calculation of Time to Ignitions

In common thermal ignition theory a material (solid or liquid) is assumed to ignite when the surface reaches the ignition temperature. It may be at the piloted or at the auto-ignition temperature. The time it takes the surface to reach such a critical temperature when heated depends on the dimensions and the thermal properties of the material. Below the special cases of thin and semi-infinite solids will be outlined in Sects. 8.2.1 and 8.2.2, respectively, as developed by Wickström [25].

In both cases the heat transfer by radiation and convection to an exposed surface is calculated according to Eq. 4.12 as

$$\dot{q}''_{tot} = \varepsilon\left(\dot{q}''_{inc} - \sigma T_s^4\right) + h_c\left(T_g - T_s\right) \tag{8.1}$$

where ε is the surface emissivity and absorptivity coefficient, σ the Stefan–Boltzmann constant, h the convection heat transfer coefficient, and T_g the ambient gas temperature. By calculating the surface temperature vs. time the time to ignition can be obtained. Eq. 8.1 is, however, a non-linear boundary condition since the emitted radiation term depends on the surface temperature to the fourth power. Therefore a direct closed form solution cannot in general be derived for the surface temperature T_s. Therefore the time to ignition must be calculated numerically.

However, for the ideal case, as, for example, in the Cone calorimeter, see Fig. 8.1, when the following conditions are present

- Constant incident radiation $\dot{q}''_{inc}$
- Constant surrounding gas temperature T_g
- Uniform initial temperature T_i
- Constant material and heat transfer properties

The time to reach the ignition temperature t_{ig} may be calculated approximately with a simple explicit formula as introduced below. Similar conditions can be assumed when analysing, for example, heating by radiation by flames or hot objects onto surfaces surrounded by gases with moderate temperature.

The formulas derived are semi-empirical, i.e. they have been proven correct by comparing with accurate numerical solutions. As a first step the third kind of BC (see Sect. 1.1.3) according to Eq. 8.1 is replaced by a second kind of BC, a constant effective heat flux $\dot{q}''_{tot,eff}$ assumed to be

$$\dot{q}''_{tot,eff} = \varepsilon\dot{q}''_{inc} - \eta\dot{q}''_{inc,cr} \tag{8.2}$$

where $\dot{q}''_{inc,cr}$ is *critical incident radiation heat flux*, i.e. the incident radiation required to balance the heat losses at the surface by emitted radiation and convection at the ignition temperature.

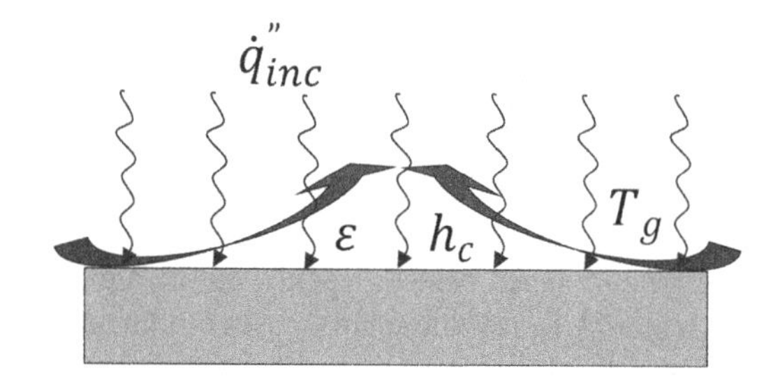

Fig. 8.1 Surface with heat transfer parameters $\boldsymbol{\varepsilon}$ and h_c exposed to constant uniform incident radiation and constant surrounding gas temperature

$$\varepsilon \dot{q}''_{inc,cr} - \varepsilon\sigma T_{ig}^4 - h_c(T_{ig} - T_g) = 0 \tag{8.3}$$

which yields

$$\dot{q}''_{inc,cr} = \sigma T_{ig}^4 + h_c/\varepsilon \cdot (T_{ig} - T_g) \tag{8.4}$$

η in Eq. 8.2 is a semi-empirical reduction coefficient which was determined by comparisons of times to ignition obtained with accurate numerical methods. The constant heat flux according to Eq. 8.2 can now be calculated for thin solids with Eq. 3.30 with $\eta = 0.3$ and for semi-infinite solids with Eq. 3.5 with $\eta = 0.8$. Then the time to ignition t_{ig} can be calculated by closed form simple equations according to Eq. 8.7 and Eq. 8.9 for thin and semi-infinite solids, respectively.

8.2.1 Thin Solids

For *thin solids* the temperature may be assumed uniform throughout the depth of the body. Then the thickness and the volumetric specific heat capacity are decisive for the time to ignition and when assuming a constant total heat flux $\dot{q}''_{tot}$ (see also Sect. 3.1) the temperature rise can be calculated as:

$$T_s - T_i = \frac{\dot{q}''_{tot} \cdot t}{\rho \cdot c \cdot d} \tag{8.5}$$

where T_s is the exposed body temperature, T_i the initial temperature, t time, ρ density, c specific heat capacity and d thickness. Density times thickness ($\rho.d$) is weight per unit area. That means that time to ignition t_{ig} of a thermally thin material receiving a *constant* total heat flux $\dot{q}''_{tot}$ is directly proportional to the density and the thickness of the material, i.e. the weight of the solid per unit area, and the temperature rise to reach the ignition temperature T_{ig}, i.e.

$$t_{ig} = \frac{\rho c d}{\dot{q}''_{tot}} (T_{ig} - T_i) \tag{8.6}$$

However, the heat flux $\dot{q}''_{tot}$ can rarely be assumed constant or determined a priori as a second kind of boundary condition. Even if exposed to a constant thermal exposure, $\dot{q}''_{tot}$ decreases when the surface temperature rises according to Eq. 8.1. As a matter of fact, it is a third kind of boundary condition, see Sect. 1.1.3. Nevertheless by inserting the effective heat flux according to Eq. 8.2 with $\eta = 0.3$ into Eq. 8.6, the time to reach the ignition temperature of thin solids exposed to radiation may be approximated as

$$t_{ig} \approx \frac{\rho \cdot c \cdot d}{\varepsilon \cdot \dot{q}''_{inc} - 0.3 \cdot \dot{q}''_{inc,cr}} (T_{ig} - T_i) \tag{8.7}$$

Calculated ignition times according to Eq. 8.7 yields very good approximations of the accurate predictions obtained by the numerical solutions where the real boundary heat flux $\dot{q}''_{tot}$, according to Eq. 8.1, is assumed and where the time to ignition is the time when the body surface reaches the ignition temperature.

Example 8.1 Calculate the time to ignition for a thin curtain with an area density of $\rho \cdot d = 300\,\text{g/m}^2$ and an ignition temperature $T_{ig} = 350\,°\text{C}$ when exposed to an incident radiation $\dot{q}''_{inc} = 20\,\text{kW/m}^2$ from both sides. Assume a specific heat capacity of curtain $c = 850$ W s/(kg K), a surrounding temperature and an initial temperature T_i equal to 20 °C, a convective heat transfer coefficient of 5 W/(m^2 K) and an emissivity $\varepsilon = 0.9$.

Solution Equation 8.7 yields $t_{ig} = \frac{850 \cdot 0.3 \cdot (350-20)}{2 \cdot (0.9 \cdot 20000 - 0.3 \cdot 10375)} = 2.8$ s.

8.2.2 Semi-infinite Solids

A similar expression as given by Eq. 8.5 for thin solids can be derived for semi-infinite solids or thermally thick solids, i.e. the thickness is larger than the thermal penetration depth, see Sect. 3.2.1. Then for a *constant* heat flux to the surface $\dot{q}''_s$ (2nd kind of BC) and constant thermal properties, the time to reach a given temperature T_{ig} is (see also Eq. 3.30)

$$t_{ig} = \frac{\pi \cdot k \cdot \rho \cdot c}{4\,(\dot{q}''_s)^2} (T_{ig} - T_i)^2 \tag{8.8}$$

where T_i is the initial temperature. The product of the heat conductivity k, the specific heat capacity c and the density ρ is the *thermal inertia* $(k \cdot \rho \cdot c)$ of the material as defined in Sect. 3.2. As for thin solids the heat flux to the surface cannot be specified as a second kind of boundary condition. However, for the particular case of a constant incident radiation flux $\dot{q}''_{inc}$ and a constant ambient gas temperature T_g and a uniform initial temperature T_i, the time to ignition can be approximated as shown below for semi-infinite solids.

By inserting the effective average value of the heat flux according to Eq. 8.2 with $\eta = 0.8$ into Eq. 8.9 and rearranging, the time to reach the ignition temperature of surfaces of solids exposed to radiation may then be approximated as

$$t_{ig} = \frac{\pi\,(k \cdot \rho \cdot c)}{4\,\varepsilon^2}\left[\frac{(T_{ig} - T_i)}{\left(\dot{q}''_{inc} - 0.8 \cdot \dot{q}''_{inc,cr}\right)}\right]^2 \tag{8.9}$$

and after inserting $\dot{q}''_{inc,cr}$ from Eq. 8.4 a closed form explicit expression for the time to ignition is obtained as

$$t_{ig} = \frac{\pi\,(k \cdot \rho \cdot c)}{4\,\varepsilon^2}\left[\frac{T_{ig} - T_i}{\dot{q}''_{inc} - 0.8\left[\sigma \cdot T_{ig}^4 + h_c\left(T_{ig} - T_g\right)\right]}\right]^2 \tag{8.10}$$

Equations 8.9 and 8.10 match very well the times to ignition as calculated by accurate numerical procedures for a wide range of the parameters incident radiation $\dot{q}''_{inc}$, ignition temperature T_{ig} and thermal inertia $(k \cdot \rho \cdot c)$ [24].

According to the above equations the inverse of the square root of the ignition time is linearly dependent on the incident radiation, i.e.

$$\frac{1}{\sqrt{t_{ig}}} = \frac{2\,\varepsilon}{\sqrt{[\pi\,(k \cdot \rho \cdot c)]}\left(T_{ig} - T_i\right)}\left[\dot{q}''_{inc} - 0.8\,\dot{q}''_{inc,cr}\right] \tag{8.11}$$

Thus according to Eq. 8.11 linear relations are obtained as shown in Fig. 8.2 for various thermal inertia and in Fig. 8.3 for various ignition temperatures. Values typical for the Cone Calorimeter test scenario has been assumed for both the diagrams in Figs. 8.2 and 8.3, i.e. initial temperature $T_i = 20\,°C$, the surface emissivity $\varepsilon = 0.9$, the convection heat transfer coefficient $h_c = 12\ W/(m^2 K)$. The lowest thermal inertia 1000 $(W^2\ s)/(m^4\ K^2)$ may be representative of low density polymeric insulation material which heats up very quickly while an inertia of 100000 $(W^2\ s)/(m^4\ K^2)$ may represent soft wood and 300000 $(W^2\ s)/(m^4\ K^2)$ hard wood such as oak. These values are only indicative and are not recommended to be used in real application. Notice that the graphs cross the abscissa at 80 % of the critical incident flux, i.e. at 0.8 $\dot{q}''_{inc,cr}$, independently of the thermal inertia of the material.

The theory indicates how significant the thermal inertia is for the time to ignition. As an example Fig. 8.2 indicates the time to ignition for softwood exposed to 30 kW/m^2 is $1/0.25^2 = 16\,s$ and $1/0.15^2 = 44\,s$ for hardwood when exposed to 30 kW/m^2. The very short ignition times for low density insulation materials even at

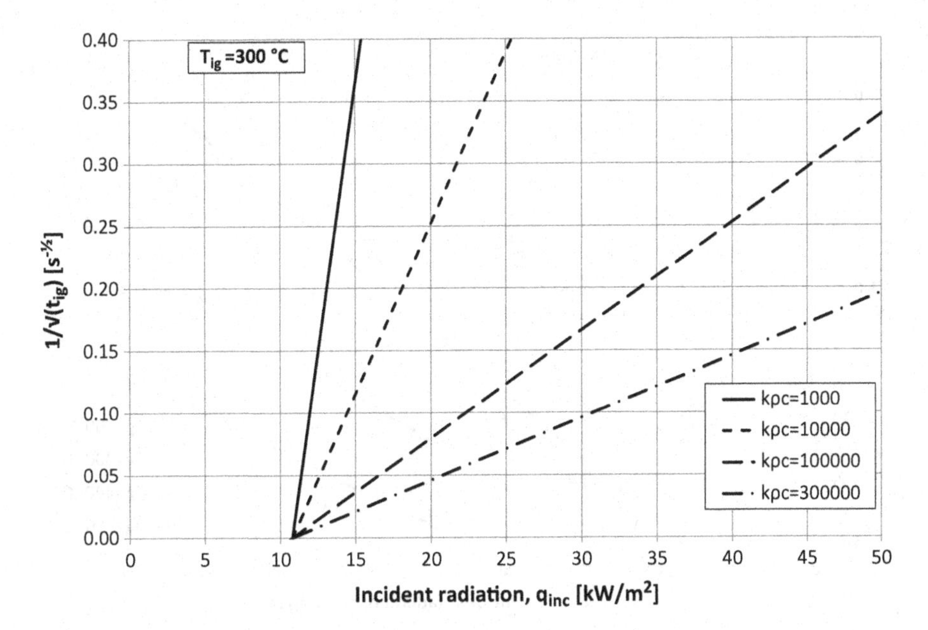

Fig. 8.2 The inverse of the square root of time to ignition vs. incident radiation heat flux according to Eq. 8.11 assuming an ignition temperature $T_{ig} = 300\,°\mathrm{C}$ for various thermal inertia $k \cdot \rho \cdot c$ given in ($\mathrm{W^2\ s)/(m^4\ K^2}$). $T_i = 20\,°\mathrm{C}$, $\varepsilon = 0.9$ and $h_{con} = 12\,\mathrm{W/(m^2K)}$

moderate incident radiation levels indicates the hazardous fire properties of these type of materials.

In Fig. 8.3 the inverse of the square root of the ignition time vs. incident radiation is shown for various ignition temperatures assuming a thermal inertia of 100000 ($\mathrm{W^2\ s)/(m^4\ K^2}$) corresponding to soft wood. The ignition temperature of 500 °C is only relevant for auto-ignition circumstances while the other temperature levels may be relevant for piloted ignitions for most materials of interest. Notice that the ignition temperature has a great influence on time to ignition and on the critical incident radiant heat flux. As an example softwood exposed to an incident radiation of 20 $\mathrm{kW/m^2}$ would ignite after 51 s when the ignition temperature is assumed to be 200 °C and after only 13 s if assumed to be 300 °C.

Example 8.2 Calculate the time to ignition of a surface of thick wood ($k \cdot \rho \cdot c = 196000\,(\mathrm{W^2\ s)/(m^4\ K)}$) solid suddenly exposed to an incident radiation heat flux $\dot{q}''_{inc} = 30\,\mathrm{kW/m^2}$. The wood surface emissivity $\varepsilon = 0.8$, the convection heat transfer coefficient $h = 12\ \mathrm{W/(m^2\ K)}$, the other thermal properties according to Table 1.2.

(a) The solid is initially at 20 °C and surrounded by air at the same temperature.
(b) The solid is initially at 100 °C and surrounded by air at the same temperature.

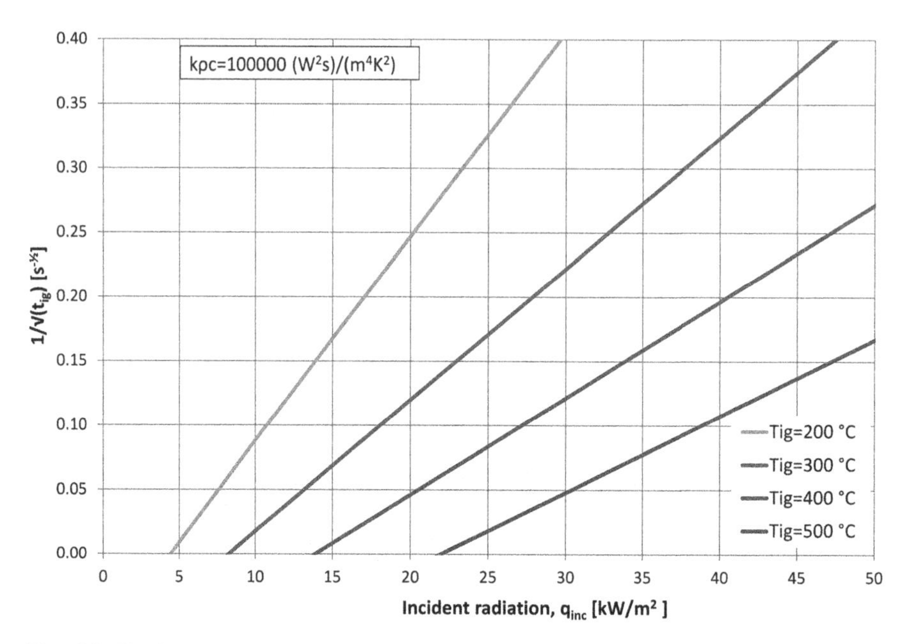

Fig. 8.3 The inverse of the square root of time to ignition vs. incident radiation heat flux according to Eq. 8.11 assuming a thermal inertia of 100000 (W2 s)/(m^4 K^2) for various ignition temperatures. $T_i = 20\,°\mathrm{C}$, $\boldsymbol{\varepsilon} = 0.9$ and $h_{\mathrm{con}} = 12\,\mathrm{W/(m^2K)}$

Solution According to Eq. 8.4 $\dot{q}''_{inc,cr} = 5.67 \cdot 10^{-8} \cdot 623^4 + 12/0.8 \cdot (350 - 20) = 13492\,\mathrm{W/m^2}$. Equation 8.9 yields $t_{ig} = \frac{\pi \cdot 196000}{4 \cdot 0.8^2} \left[\frac{350-20}{(30000-0.8 \cdot 13492)}\right]^2 = 71\,\mathrm{s}$. (*Comment*: Applying the accurate boundary condition according to Eq. 8.1 yields by accurate numerical calculations $t_{ig} = 69\,\mathrm{s}$).

According to Eq. 8.4 $\dot{q}''_{inc,cr} = 5.67 \cdot 10^{-8} \cdot 623^4 + 12/0.8 \cdot (350 - 100) = 12292\,\mathrm{W/m^2}$. Equation 8.9 yields $t_{ig} = \frac{\pi \cdot 196000}{4 \cdot 0.8^2} \left[\frac{350-100}{(30000-0.8 \cdot 12402)}\right]^2 = 37\,\mathrm{s}$.

Comment: Thus this material would ignite in about half the time if preheated from 20 to 100 °C.

Chapter 9
Measurements of Temperature and Heat Flux

In FSE temperature is nearly always measured with thermocouples as described in Sect. 9.1. Heat flux measured in different ways is most commonly measured as the sum of the net heat flux by radiation and convection to a cooled surface. The principles are briefly outlined in Sect. 9.2. Alternative methods incident radiation heat flux as well AST using so-called plate thermometers has also been developed as a practical alternative to heat flux meters as outlined in Sect. 9.3.

9.1 Thermocouples

Thermocouples (sometimes abbreviated T/C) have a junction between a pair of wires of two different metals or alloys. A voltage is then generated proportional to the temperature difference between the so-called hot junction and the cold junction, a reference point with known temperature. The hot junction of thermocouples can either be imbedded in solid materials or mounted in free space. Thermocouples are in general relatively inexpensive and easy to handle, and can be used for measuring temperatures over a wide range. They are therefore very common in fire testing and research. Different alloys are used for different temperature ranges.

9.1.1 Type of Thermocouples

There are a number of standardized types and combinations metals for thermocouples. The most common have been designated letters by ISA (Instrument Society of America) and ANSI (American National Standards Institute). Information on the various types of thermocouples and their letter designation is given in the international standard IEC 584.

U. Wickström, *Temperature Calculation in Fire Safety Engineering*,
DOI 10.1007/978-3-319-30172-3_9

In fire testing and research thermocouples of type K are by far the most common. The positive lead is then made of a nickel alloy with 10 % chrome and the negative of a nickel alloy with 2 % aluminium, 2 % mangan and 1 % silicon. The relation between the output voltage and temperature is almost linear with a sensitivity of approximately 41 μV/K. The melting point is about 1400 °C and the mechanical properties and the resistance against corrosion are satisfactory also at high temperature levels. At temperatures above 800 °C, however, oxidation may occur leading to substantial measuring errors. The thermocouples may also age when used for longer times at temperatures above 500 °C and should therefore in such cases be calibrated about every 20 h of use [26]. According to the international standards for fire resistance furnace tests ISO 834 and EN 1363-1 thermocouples may not be used for more than 50 tests.

For temperature measurements up to 1480 °C thermocouples of type S, platinum—platinum/rhodium, are sometimes used in fire resistance furnaces. They are, however, expensive and only suitable for short term measurements as they degrade at high temperatures.

There are industrially manufactured thermocouples in many metal combinations. The connections of the thermocouples leads are, however, often made by the user by soldering, electric or gas welding or pressing depending on the intended use. Soldering and welding is mainly used for thin thermocouples. Pressing is used with so-called quick tips. The latter method yields rather big hot junctions which makes the thermocouples relatively slow when recording dynamic processes as in fires and less sensitivity to heat transfer by convection, see next Section. Thus their temperature recordings must often be corrected due to the effects of radiation when measuring gas temperature accurately.

Ready-made *shielded thermocouples* are also being used in fire testing and research. They have a stainless steel or similar casing protecting the thermocouple from mechanical stresses and corrosive gases. These are in general more robust but considerably more costly.

9.1.2 Measurement of Temperature in Gases

The temperature recording you get from a thermocouple is always the temperature of the hot junction of the thermocouple leads. When placed in a gas it adjusts more or less quickly to surrounding temperatures depending on its thermal response characteristics to convection and radiation. Briefly it can be said that the smaller dimensions of a thermocouple the quicker it responses to thermal changes and the more sensitive it is to convection and thereby gas temperature, and vice versa.

Thus it is important to realize that thermocouples in gases are influenced by the gas temperature T_g as well as by the incident radiation or the black body radiation temperature T_r (see Chap. 4). It adjusts to a temperature which is a weighted average of the two temperatures which may be very different. The weights are the heat transfer coefficients h_r and h_c for radiation and convection, respectively. Thus the ultimate or equilibrium thermocouple temperature T_{TC} becomes

$$T_{TC} = \frac{h_r \cdot T_r + h_c \cdot T_g}{h_r + h_c} \tag{9.1}$$

where

$$h_r = \varepsilon\, \sigma \left(T_r^2 + T_{\mathrm{TC}}^2\right) \left(T_{\mathrm{r}} + T_{TC}\right) \tag{9.2}$$

Notice that Eq. 9.1 is *implicit* as h_r depends on T_{TC}, and that it is similar to the expression for the AST in Eq. 4.23. When radiation and thermocouple temperatures are approximately equal, i.e. $T_{\mathrm{r}} \approx T_{TC}$ (as for thermocouples in thick flames) the radiation heat transfer coefficient may be approximated as

$$h_r \approx 4\varepsilon \cdot \sigma \cdot T_{\mathrm{TC}}^3 \tag{9.3}$$

As indicated in Chap. 6, convection heat transfer coefficients decrease with the size of a body. Hence smaller thermocouples will have greater convection heat transfer coefficients h_c and will therefore adjust closer to the gas temperature while larger thermocouples will deviate more from the gas temperature and adjust closer to the radiation temperature as indicated by Eq. 9.1.

The difference between the true gas temperature and the thermocouple temperature at equilibrium can be written as

$$\Delta T = T_{TC} - T_g = \frac{h_r}{h_c} \left(T_{\mathrm{r}} - T_{\mathrm{TC}}\right) \tag{9.4}$$

which implies that the difference between the measured temperature T_{TC} and the true gas temperature T_g increases with the ratio between the heat transfer coefficients and the difference between the radiation temperature and the thermocouple temperature.

A special case is the plate thermometer as described in Sect. 9.3 which has a large exposed area. The convection heat transfer coefficient h_c is therefore relatively small, and hence the equilibrium temperature of a PT is closer to the incident radiation temperature than the corresponding temperature of ordinary thermocouples. In addition the PT is dependent on direction of incident radiation while a thermocouple is not.

The time response characteristics are also important to consider when measuring gas temperatures. A general rule is thermocouples response faster the thinner they are, the less mass they have. As the temperature in a thermocouple can be assumed uniform and it can be calculated assuming lumped-heat-capacity (Sect. 3.1). Thus the temperature T_{TC} of a thermocouple suddenly exposed to a constant fire temperature T_f may be calculated according to Eq. 9.5 as

$$\frac{T_{TC} - T_i}{T_f - T_i} = 1 - e^{-\frac{t}{\tau}} \tag{9.5}$$

where T_i is the initial thermocouple temperature and τ is the *time constant* of the thermocouple which then can be calculated as

$$\tau = \frac{V_{TC} \cdot \rho_{TC} \cdot c_{TC}}{A_{TC} \cdot h_{tot}} \tag{9.6}$$

where the parameters ρ_{TC} and c_{TC} are the density and the specific heat capacity of the thermocouple junction including the mass of soldering, etc. and V_{TC}/A_{TC} is the effective volume-to-area ratio. When assuming constant conditions the value time constant is the time elapsed when the temperature rise has reached 63 % of its final value. In reality the time constant varies considerably as h_r increases significantly with temperature.

For a total heat transfer (by convection and radiation) coefficient h_{tot}, the time constant τ for a *sphere* can be identified as

$$\tau = \frac{1}{6} D \cdot \rho_s \cdot c_s / h_{tot} \tag{9.7}$$

Assuming the thermocouple hot junction as a *cylinder* with a diameter D disregarding the end surface yields

$$\tau = \frac{D}{4} \rho_s \cdot c_s / h_{tot} \tag{9.8}$$

When exposure temperature T_f varies with time the thermocouple temperature can be obtained to Eq. 3.16.

Example 9.1 What is the time constant τ of a thermocouple T/C exposed to uniform temperature at a level of $T_f = 500\,°\mathrm{C}$? Assume that the *T/C* is spherical with a diameter of 3 mm of steel with a convective heat transfer coefficient $h_c = 50$ W/m^2 K. The density and specific heat capacity of the T/C may be assumed to be 7850 kg/m^3 and 460 Ws/kg K, respectively, and its emissivity $\varepsilon = 0.9$.

Guidance: *Assume the heat transfer to the T/C is* $\dot{q}'' = \varepsilon\sigma\left(\overline{T}^4{}_f - \overline{T}^4{}_{TC}\right) + h_c \left(T_f - T_{TC}\right)$ *and that* $T_{TC} = T_f$ *when calculating the radiation heat transfer coefficient* h_r.

Solution Equation 9.3 (or Fig. 4.2a) yields $h_r = 4 \cdot 0.9 \cdot 5.67 \cdot 10^{-8} \cdot (500 + 273)^3 = 105\,\mathrm{W}/(\mathrm{m}^2\,\mathrm{K})$ and then $h_{tot} = 105 + 50 = 155\ \mathrm{W}/(\mathrm{m}^2\,\mathrm{K})$. According to Eq. 9.7 $\tau = \frac{1}{6} \cdot 0.003 \cdot 7850 \cdot 460/155 = 11.6\,\mathrm{s}$.

Example 9.2 Calculate the time constant of the 1 mm thick thermocouple in Example 6.3.

(a) Initially at room temperature (300 K)
(b) At its equilibrium temperature (1000 K)

when suddenly exposed to gas and radiation temperatures of 1000 K. Assume the T/C is made of stainless steel, i.e. $\varepsilon_{TC} = 0.7$, $\rho_{TC} = 7900\,\text{kg/m}^3$ and $c_{TC} = 460\,\text{W}/(\text{kg K})$.

Solution

(a) According to Example 6.3 $h_c = 131\,\text{W}/(\text{m}^2\,\text{K})$. The radiation heat transfer coefficient is obtained from Eq. 4.5 as $h_r = 0.7 \cdot 5.67 \cdot 10^{-8}(1000^2 + 300^2)(1000 + 300) = 56\,\text{W}/(\text{m}^2\,\text{K})$. Then $h_{tot} = 131 + 56 = 187\ \text{W}/(\text{m}^2\,\text{K})$ and $\tau_{TC} = 0.001/4 \cdot 7900 \cdot 460/187 = 4.8\,\text{s}$.
(b) According to Example 6.3 the convection $h_c = 147\ \text{W}/(\text{m}^2\ \text{K})$. According to Eq. 4.6 as $h_r = 4\varepsilon\sigma T_r^3 = 159\,\text{W}/(\text{m}^2\,\text{K})$ and $h_{tot} = 147 + 159 = 306\,\text{W}/(\text{m}^2\,\text{K})$. Now the time constant of the thermocouple can be estimated according to Eq. 9.8 as $\tau_{TC} = 0.001/4 \cdot 7850 \cdot\ 460/306 = 3.0\,\text{s}$.

9.1.3 Corrections of Time Delay

All thermocouples respond to the thermal exposure with a time delay depending on the thermocouple characteristics and thermal environment as described above. When the response of a thermocouple can be expressed as in Eq. 9.5, i.e. assuming lumped-heat-capacity, the value of T_f at a given time may be obtained numerically by solving the so-called inverse problem. The time constant τ must either be known explicitly or implicitly, for example, as functions of the exposure temperature T_f and the response temperature T_{TC} as shown by Eq. 9.7. The time derivative of the thermocouple temperature may then at any arbitrary time t be derived from Eq. 9.5 as

$$\frac{dT_{TC}}{dt} = \frac{T_f - T_{TC}}{\tau} \tag{9.9}$$

Given a series of thermocouple recordings true exposure level T_f can be derived from Eq. 9.9. The time derivative of the thermocouple temperature is then approximated by the corresponding differential between two consecutive thermocouple recordings and the following expression can be derived:

$$T_f^{j+1} = \left(1 + \frac{\tau}{\Delta t}\right) T_{TC}^{j+1} - \frac{\tau}{\Delta t} T_{TC}^{j} \tag{9.10}$$

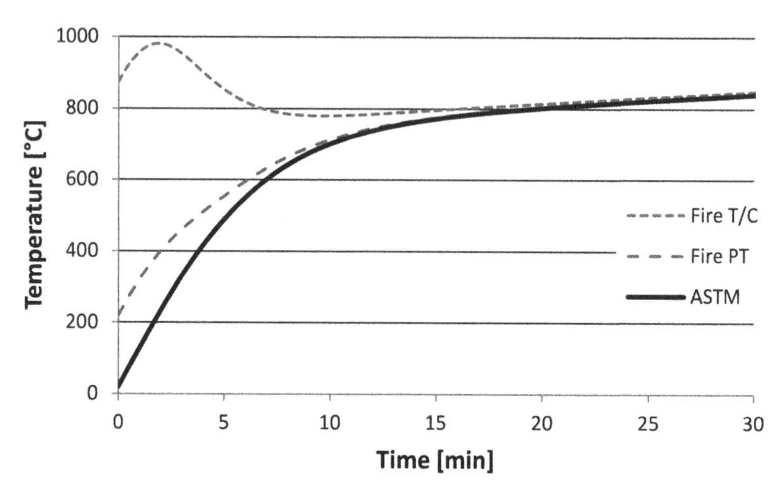

Fig. 9.1 The ASTM-E119 fire curve and temperatures the thermocouple according to ASTM E-119 and the PT according to ISO 834 must follow to obtain the specified furnace temperature due to time delay

where Δt is a time increment of the measurement and j is the measurement number. The accuracy and the numerical stability of such a calculation depends on the relation between τ and Δt.

As an example Fig. 9.1 shows the actual furnace temperature rise in a furnace controlled ideally precisely according to the ASTM E-119 standard fire curve with temperature monitoring thermocouples according to the standard time constants in the range of 5–7.2 min. Notice that the real or effective furnace temperature is much higher than indicated by the slowly responding ASTM type of shielded thermocouples. The diagram also shows for comparison the corresponding curve for a standard PT according to ISO 834. This curve does not deviate as much from the ideal standard ASTM time–temperature curve as the time constant of a PT is much shorter than the time constant of an ASTM thermocouple.

This implies that when predicting temperature in specimens being exposed to a standard ASTM E-119 furnace test it is important to assume a much higher exposure temperature for the first 10 min than what has been recorded in the test by the standard thermocouples. In calculations for deriving the T/C and PT response curves in Fig. 9.1 properties according to Table 9.1 were applied. These values are reasonable but uncertain depending on among other thing furnace characteristics and therefore the curves of Fig. 9.1 should just be taken as indicative implying that the influence of the time delay is significant for an ASTM-E119 test and must be considered in particular when predicting temperature in structures exposed to short test durations.

Table 9.1 Parameters used for analysing the time delay of the ASTM-E119 thermocouple and the ISO 834 PT in combination with Eq. 4.12

	Effective thickness, d [mm]	Convection heat transfer coefficient, h_c [W/(m^2 K)]	Emissivity, ε [-]
ASTM-E119 thermocouple	6	50	0.8
ISO 834 plate thermometer	0.7	25	0.8

9.2 Heat Flux Meters

In several fire test methods incident radiation levels are specified. Therefore, it is important that radiation heat flux can be measured with sufficient accuracy. It is usually measured with so-called total heat flux meters of the Gardon or Schmidt-Boelter types. Such meters register the combined heat flux by radiation and convection to a water cooled surface. Thus the measurement will contain contributions by convection which depends on a number of factors such as the design of the heat flux meter, the orientation of the meter, the cooling water temperature, the local temperature and gas/air flow conditions. In unfavourable conditions the uncertainty due to convection can amount to 25 % of the total heat flux, see ISO 14934. As a general rule the error is lesser when the meter is surrounded by a gas at a temperature close to the cooling water temperature while the errors may be very large when the meter is exposed to hot fire gases or flames. Under such conditions Gardon or Schmidt-Boelter type meters are both impractical and inaccurate. Then devices such as the PT as described in Sect. 9.3 are more useful.

The principal designs of a Gardon and a Schmidt-Boelter heat flux meters are shown in Fig. 9.2. In the Gardon gauge the temperature difference between the middle of the circular disc and its water cooled periphery is proportional to the received heat flux by radiation and convection. In the Schmidt-Boelter gauge the temperature difference between the exposed surface and a point at a depth below is measured with a so-called thermopile including several hot and cold junctions. This type of HFM therefore yields a higher output voltage than a Gardon gauge for the same flux.

9.2.1 Calibration and Use of Heat Flux Meters

Heat flux meters such as the Gardon gauge and Schmidt-Boelter gauge are calibrated according to ISO 14934 in a spherical furnace with a uniform temperature. The gauge is then exposed to an incident radiation proportional to the fourth power of the furnace temperature T_{fur}^4. The heat transfer by convection is negligible in the calibration configuration (see Fig. 9.3) and therefore the heat transfer to the water cooled sensing surface is:

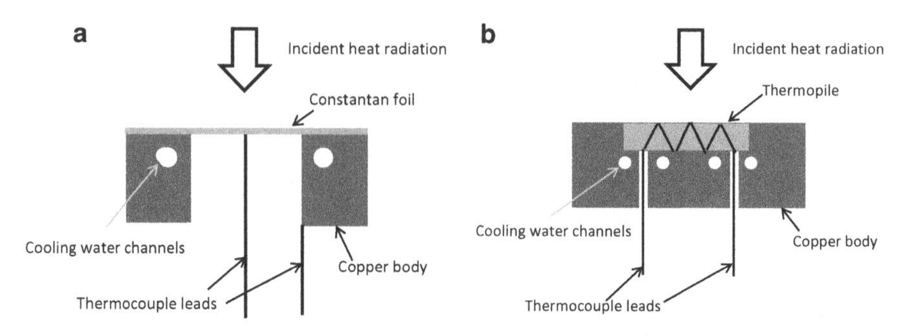

Fig. 9.2 Principal cross sections of total heat flux meters. (**a**) Gardon gauge (**b**) Schmidt-Boelter gauge

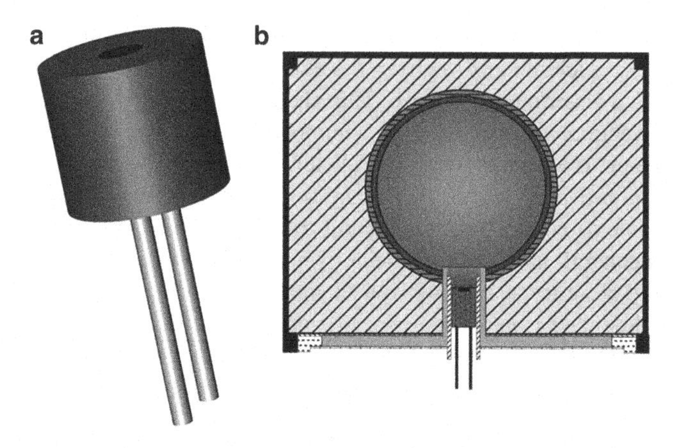

Fig. 9.3 A 3-D drawing of a heat flux meter and a spherical calibration furnace with a heat flux meter mounted in the bottom opening. (**a**) Heat flux meter (**b**) Calibration furnace

$$\dot{q}''_{hfg} = \varepsilon_{hfm}\, \sigma\left(\varphi\, T^4_{fur} - T^4_{hfm}\right) \tag{9.11}$$

where T_{hfm} is the surface temperature of the sensing body. At high furnace temperature or heat flux levels the second term of the above equation is relatively very small and can be neglected. Otherwise T_{hfm} is assumed equal to the cooling water temperature. The coefficient φ is a test configuration parameter depending on the geometric configuration when a HFM is mounted in the test furnace.

The calibration procedure of a HFM means that the electric voltage output is determined at several heat flux levels obtained by various furnace temperature levels. Then a normally linear relation can be established between the heat flux $\dot{q}''_{hfm}$ and the output voltage.

When in use and exposed to radiation and convection the general expression of the heat flux $\dot{q}''_{hfm}$ to the sensing body of a HFM is:

$$\dot{q}''_{hfm} = \varepsilon_{hfm}\left(\dot{q}''_{inc} - \sigma T^4_{hfm}\right) + h_{hfm}\left(T_g - T_{hfm}\right) \tag{9.12}$$

where T_g is the gas temperature near the HFM. The incident heat flux $\dot{q}''_{inc}$ can be obtained, given the gas temperature T_g and the emissivity ε_{hfm} and the convection heat transfer coefficient h_{hfm} are known, as

$$\dot{q}''_{inc} = \frac{1}{\varepsilon_{hfm}}\left[\dot{q}''_{hfm} - h_{hfm}\left(T_g - T_{hfm}\right)\right] + \sigma T^4_{hfm} \tag{9.13}$$

Often it is assumed that $\varepsilon_{hfm} \approx 1$ and the term $\sigma\ T^4_{hfm}$ is negligible, and when $T_g \approx T_{hfm}$, the convection term vanishes as when placed in air at ambient temperature. Then the incident radiation heat flux can be approximated as

$$\dot{q}''_{inc} = \dot{q}''_{hfm} \tag{9.14}$$

When then using the measured data for calculating the heat transfer to a target surface with a temperature T_s based on HFM measurements the general expression according to Eq. 4.11 applies, and the total heat flux $\dot{q}''_{tot}$ to a surface can be derived by inserting Eq. 9.13 into Eq. 4.11:

$$\begin{aligned}\dot{q}''_{tot} = &\frac{\varepsilon_s}{\varepsilon_{hfm}}\left[\dot{q}''_{hfm} + \varepsilon_{hfm}\sigma T^4_{hfm} - h_{hfm}\left(T_g - T_{hfm}\right)\right] \\ &-\varepsilon_s\sigma T^4_s + h_c\left(T_g - T_s\right)\end{aligned} \tag{9.15}$$

Thus the total heat transfer depends on the emissivity and the convection heat transfer coefficient of both the HFM and the target surface. These parameters are often not very well known which introduces great uncertainties especially when the HFM is placed in hot gases or flames with temperatures deviating from the cooling water temperature. Then the uncertainty due to the convection becomes significant as the heat transfer by convection to a HFM with its small surface is difficult to estimate accurately. However, usually the emissivities of the HFM and the target surface are assumed equal, and when the gas and cooling water temperature are assumed equal as well, then the heat transfer to an adjacent target surface becomes independent of the gas temperature T_g and the expression of the total heat flux becomes:

$$\dot{q}''_{tot} = \dot{q}''_{hfm} + \varepsilon \cdot \sigma T^4_{hfm} - \varepsilon \cdot \sigma \cdot T^4_s - h\left(T_s - T_{hfm}\right) \tag{9.16}$$

There are several uncertainties in this expression. A more complete analysis of the use of heat flux meters are given by Lattimer [27].

Example 9.3 A water cooled heat flux meter is used to measure the total incident heat flux from a fire against a wall painted black. The measured heat flux is 30 kW/m^2 and the water cooled gauge is measured to be 350 K. Both the wall emissivity and the heat flux gauge have a surface emissivity $\varepsilon = 0.95$, and the heat transfer coefficient is 10 W/m^2 K.

(a) Determine the total heat flux $\dot{q}''_{tot}$ into the wall when its surface temperature is 600 K, and 800 K.
(b) Given the gas temperature T_g is 300 and 1000 K, respectively, what is the incident radiant heat $\dot{q}''_{inc}$?
(c) What is the AST, i.e. the temperature of the surface when the net heat flux into the wall vanishes, for the two gas temperature levels?
(d) Use the ASTs calculated in (c) to calculate the net heat fluxes to the surfaces when the surface temperature is 600 K and 800 K, respectively. Compare with the results obtained in (a).

Solutions

(a) Equation 9.16 yields: $\dot{q}''_{tot} = 30000 - 10 \cdot (T_s - 350) - 0.95 \cdot 5.67 \cdot 10^{-8} \cdot (T_s^4 - 350^4)$. Then for $T_s = 600$ the total heat flux $\dot{q}''_{tot} = 21.3 \cdot 10^3$ W/m^2, and for $T_s = 800$ K the total heat flux $\dot{q}''_{tot} = 4.2 \cdot 10^3$ W/m^2.
(b) For $T_g = 300\,\text{K}$, then according to Eq. 9.13 $\dot{q}''_{inc} = \frac{1}{0.95} \cdot [30000 - 10 \cdot (300 - 350)] + 5.67 \cdot 10^{-8} \cdot 350^4 = 30500 + 851 = 32.96 \cdot 10^3\,\text{W/m}^2$. For $T_g = 1000\,\text{K}$ then $\dot{q}''_{inc} = \frac{1}{0.95}[30000 - 10(1000 - 350)] + 5.67 \cdot 10^{-8} \cdot 350^4 = 24740 + 851 = 25.59 \cdot 10^3\,\text{W/m}^2$.
 Comment: Notice that the incident radiation may be considerably different for the same heat flux meter recordings. The heat transfer coefficient is taken from Lattimer's [28]. It may be considerably higher in reality which would enhance the differences.
(c) For $T_g = 300\,\text{K}$, then according to Eq. 4.21 $\varepsilon \cdot (\dot{q}''_{inc} - \sigma T_{AST}^4) + h_c(T_g - T_{AST}) = 0.95 \cdot [32.96 \cdot 10^3 - (5.67 \cdot 10^{-8}) \cdot T_{AST}^4] + 10 \cdot (300 - T_{AST}) = 0$. Thus by iteration $T_{AST} = 833\,\text{K}$. For $T_g = 1000\,\text{K}$, then $0.95 \cdot [25.59 \cdot 10^3 - (5.67 \cdot 10^{-8}) \cdot T_{AST}^4] + 10 \cdot (1000 - T_{AST}) = 0$. Thus $T_{AST} = 833\,\text{K}$.
(d) According to Eq. 4.31 $\dot{q}''_{tot} = \varepsilon \cdot \sigma(T_{AST}^4 - T_s^4) + h_c(T_{AST} - T_s)$. Then for $T_s = 600\,\text{K}$ $\dot{q}''_{tot} = 0.95 \cdot (5.67 \cdot 10^{-8}) \cdot (833^4 - 600^4) + 10 \cdot (833 - 600) = 21.3 \cdot 10^3\,\text{W/m}^2$, and for $T_s = 800\,\text{K}$ $\dot{q}''_{tot} = 0.95 \cdot 5.67 \cdot 10^{-8} \cdot (833^4 - 800^4) + 10 \cdot (833 - 800) = 4.2 \cdot 10^3\text{W/m}^2$.

 Comment: Exactly the same values were obtained when calculating the total heat flux based on $\dot{q}''_{hfm}$ as based on T_{AST} according to the theory presented. In practice it would most certainly be more expedient to use PTs for measuring ASTs and use those measurements for calculating heat flux and temperature of the exposed wall.

9.3 The Plate Thermometer

9.3.1 Introduction

The standard Plate Thermometer PT as specified in the international ISO 834 and in the European EN 1363-1 was invented to measure and control temperature in fire European resistance furnaces [27] with the purpose of harmonizing the thermal exposure and assuring tests results independent of type of fuel and furnace design.

The standard PT as shown in Fig. 9.4 is made of a shielded thermocouple attached to the centre of a 0.7-mm-thick metal plate of Inconel 600 (a trade name of an austenitic nickel based super alloy for high temperature oxidation resistance) which is insulated on its back side. The exposed front face is 100 mm by 100 mm. The back side insulation pad is 10 mm thick.

A relatively large sensor surface, such as a PT, measures neither the gas temperature nor the incident radiation or radiation temperature but a temperature between the radiation and gas temperatures. It measures approximately the temperature of a surface which cannot absorb any heat. This temperature has been named the *Adiabatic Surface Temperature*, *AST* [29–31], see Sect. 4.4. PT can also be used to measure incident radiant heat flux to a surface [32, 33] as will be shown in the next section.

As shown in Chap. 4, the concept of the AST is very valuable as it can be used for calculation of heat transfer to fire-exposed body surfaces when exposed to convection and radiation boundary conditions, so-called mixed boundary conditions, where the gas temperature and the radiation temperatures may be considerably different. Figure 9.5 shows PTs being mounted in different directions around a steel girder.

The concept of AST is not limited to fire resistance scenarios and predictions of structural element temperatures. It can also be used at more moderate temperature levels for instance to estimate whether a surface will reach its ignition temperature when exposed to elevated incident radiation but moderate gas temperatures.

9.3.2 Theory for Measuring Incident Heat Flux and Adiabatic Surface Temperature with Plate Thermometers

A simplified heat balance equation of the exposed surface plate of a PT may be written in one dimension as [33] (see also Fig. 9.6)

$$\varepsilon_{PT}\dot{q}''_{inc} - \varepsilon_{PT}\sigma T_{PT}^4 + h_{PT}(T_g - T_{PT}) + K_{PT}(T_g - T_{PT}) = C_{PT}\frac{dT_{PT}}{dt} \quad (9.17)$$

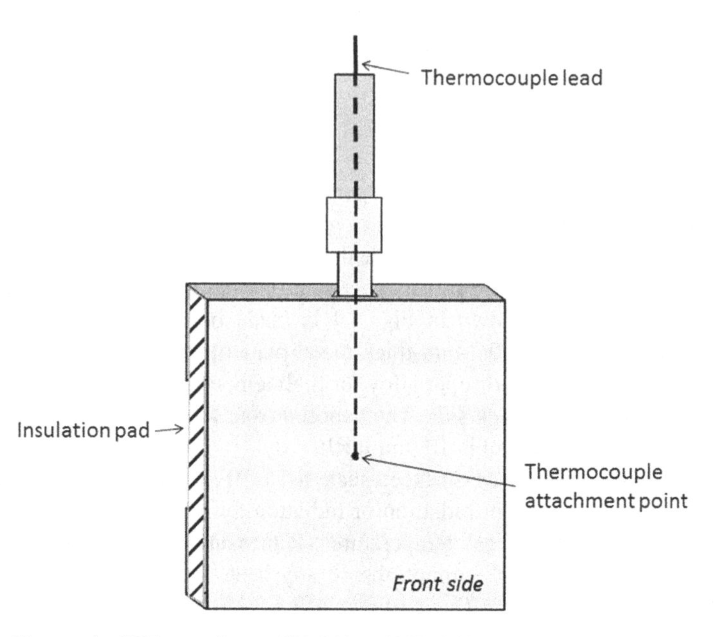

Fig. 9.4 The standard PT according to ISO 834 and EN 1363-1

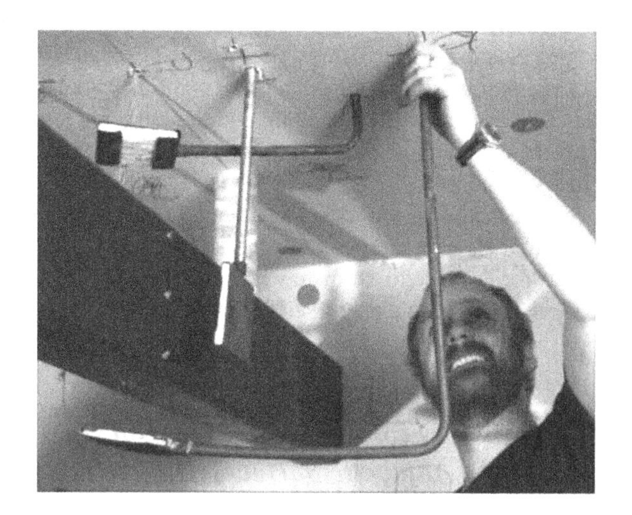

Fig. 9.5 PTs being mounted for measuring ASTs in different directions at the surfaces around a steel girder [29]

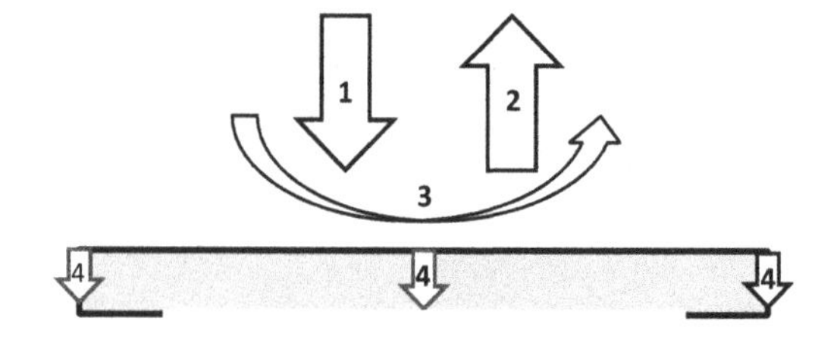

Fig. 9.6 Indication of the heat transfer to a PT. The numbers relate to the terms on the *left-hand side* of Eq. 9.17

The first term on the left-hand side of Eq. 9.12 is the radiant heat absorbed by the Inconel plate, the second the heat emitted; the third the heat transferred by convection and the fourth term expresses the heat lost by conduction through the insulation pad plus along the Inconel plate. The latter is assumed to be proportional to the difference between the plate temperature T_{PT} and the gas temperature T_g with the proportionality constant denoted K_{PT}.

The term on the right-hand side of the equation is the rate of heat stored calculated assuming lumped-heat-capacity (see Sect. 3.1). C_{PT} is assumed to be the heat capacity of the Inconel plate plus a third of the heat capacity of the insulation pad. (The third is taken from experiences of insulated steel structures, see Sect. 13.3.1).

A thorough two-dimensional thermal finite element analysis of the standard ISO 834 PT is presented in [32]. It was then found that with the thermal conduction coefficient $K_{PT} = 8.0\,\mathrm{W/(m^2\,K)}$ and the heat capacity $C_{PT} = 4200\,\mathrm{J/(m^2\,K)}$ there was a good agreement between PT temperatures calculated with FE analyses and the temperatures obtained using Eq. 9.20. The convection heat transfer coefficient of the PT h_{PT} depends on the actual scenario. In the case of natural convection only, it may be assumed to be in the order of 10 W/(m^2 K), see Sect. 6.3.1.1.

The incident radiation $\dot{q}''_{inc}$ can be derived from Eq. 9.18 as

$$\dot{q}''_{inc} = \sigma T_{PT}^4 - \frac{1}{\varepsilon_{PT}}\left[(h_{PT} + K_{PT}) \cdot (T_g - T_{PT}) - C_{PT}\frac{dT_{PT}}{dt}\right] \tag{9.18}$$

The derivative of the transient term can be approximated by the differential, i.e. $\frac{\mathrm{dT_{PT}}}{\mathrm{dt}} \approx \frac{\Delta \mathrm{T_{PT}}}{\Delta \mathrm{t}}$. Then $\dot{q}''_{inc}$ can be obtained by a stepwise procedure where $\frac{\Delta \mathrm{T_{PT}}}{\Delta \mathrm{t}} = \frac{\mathrm{T_{PT}^{j+1} - T_{PT}^{j}}}{t^{j+1} - t^{j}} = \frac{\mathrm{T_{PT}^{j+1} - T_{PT}^{j}}}{\Delta \mathrm{t}}$.

Under steady state or relatively slow processes the transient term can be neglected. In addition at high incident radiation levels the first term is dominant and the dependence on conduction and convection is relatively small and may in approximative analyses even be neglected. In Fig. 9.7 the incident radiation flux $\dot{q}''_{inc}$ is shown as a function of the temperature T_{PT} of a PT mounted vertically *in air at ambient temperature*, $T_g = 20\,°\mathrm{C}$. The emissivity is assumed $\varepsilon_{PT} = 0.9$ and the natural convection heat transfer coefficient is calculated accurately as a function of temperature according to Eq. 6.30. The incident radiation flux $\dot{q}''_{inc}$ is shown with the assumption of the heat loss by conduction parameter being neglected $K_{PT} = 0$ and $K_{PT} = 4\,\mathrm{W/(m^2K)}$, respectively. The latter is representative for a so-called insPT as shown in Fig. 9.9 with a 20 mm insulation pad. As can be observed the influence of the uncertain parameter K_{PT} is relatively small in comparison to the uncertainties related to measurements with conventional heat flux meters, see Sect. 9.2.

K_{PT} and C_{PT} may often be neglected, in particular when insulated plate thermometers insPTs as shown in Fig. 9.10 are used. Then the incident heat flux can be calculated as

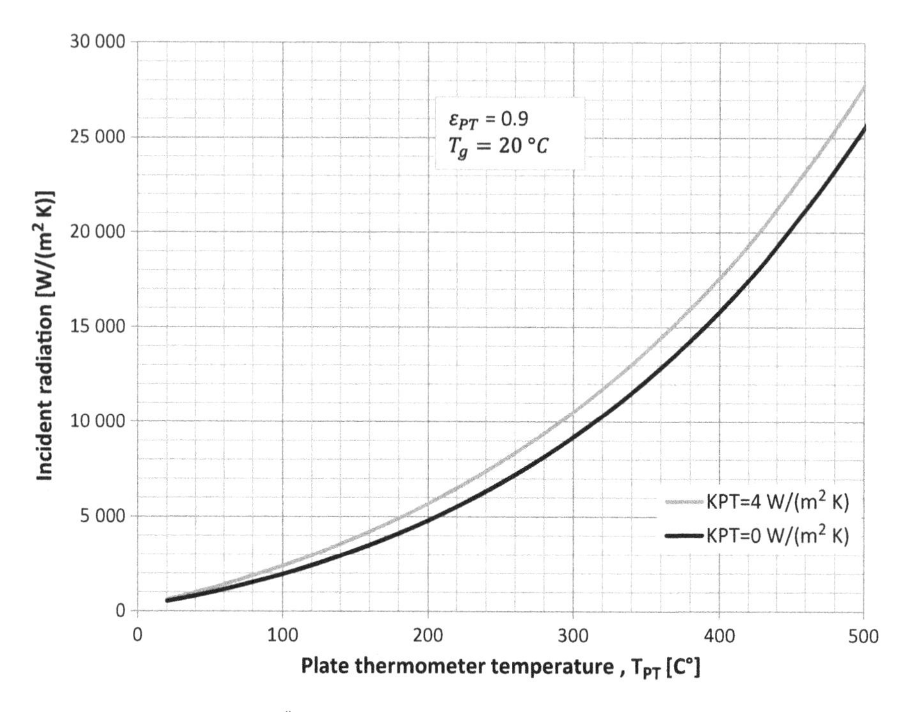

Fig. 9.7 Incident radiation $\dot{q}''_{inc}$ based on steady-state PT measurements in ambient air assuming the heat loss parameter by conduction negligible $K_{PT} = 0$ and $K_{PT} = 4\,\text{W}/(\text{m}^2\text{K})$, respectively. The heat loss by natural convection is calculated according to Eq. 6.30

$$\dot{q}''_{inc} = \sigma T_{PT}^4 - \frac{h_{PT}}{\varepsilon_{PT}}\left(T_g - T_{PT}\right) \tag{9.19}$$

The lower curve of the diagram of Fig. 9.7 shows the relation between T_{PT} and $\dot{q}''_{inc}$ when neglecting both K_{PT} and C_{PT}.

T_{AST} can be derived from PT recordings considering K_{PT} and C_{PT} by heat balance equation

$$\varepsilon_{PT}\sigma\left(T_{AST}^4 - T_{PT}^4\right) + h_{PT}(T_{AST} - T_{PT}) + K_{PT}\left(T_g - T_{PT}\right) = C_{PT}\frac{dT_{PT}}{dt} \tag{9.20}$$

Given a series of T_{PT} measurements the derivative of the transient term can be approximated by the differential $\frac{\text{dT}_{\text{PT}}}{\text{dt}} \approx \frac{\Delta\text{T}_{\text{PT}}}{\Delta\text{t}}$, and the inverse problem of calculating T_{AST} can be done by a step-by-step procedure. At each time step *j* the fourth grade equation below derived from Eq. 9.20 must then be solved

$$\varepsilon_{PT}\sigma\left(T_{AST}^{j+1}\right)^4 + h_{PT}T_{AST}^{j+1} - C_{PT}\frac{T_{PT}^{j+1} - T_{PT}^{j}}{\Delta t} - \varepsilon_{PT}\sigma\left(T_{PT}^{j+1}\right)^4 - h_{PT}T_{PT}^{j+1} + K_{PT}\left(T_g^{j+1} - T_{PT}^{j+1}\right) = 0 \quad (9.21)$$

where all parameters are known except the adiabatic surface temperature T_{AST}^{j+1}. If both K_{PT} and C_{PT} are neglected T_{AST} can be obtained at each time as from the fourth degree equation derived from Eq. 9.20

$$\varepsilon_{PT}\sigma T_{AST}^4 + h_{PT}T_{AST} - \left[\varepsilon_{PT}\sigma T_{PT}^4 + h_{PT}T_{PT}\right] = 0 \quad (9.22)$$

Solution techniques of this type of incomplete fourth degree equations are shown in Sect. 4.4.1.1.

The standard PT has successfully been used in an ad hoc test series for measuring ASTs which has then been used to predict temperature in a steel section. Figure 9.8 shows a beam near the ceiling being exposed to an intense pre-flashover fire with very uneven and complex temperature distribution.

Temperatures were then compared with measured temperatures. An example is shown in Fig. 9.9. Notice the high similarity between the measured and calculated steel temperatures. Temperatures measured with ordinary thermocouples were generally very different from those measured with PTs at similar positions. Therefore predictions of steel temperatures based on thermocouple recordings as input would not yield such good agreements between calculations and measurements. Alternative measuring techniques using, for example, heat flux meters would not have been possible as these types of instruments cannot cope with high temperature environments.

9.3.3 Alternative Plate Thermometer Designs

To achieve high accuracies it follows from an analysis of Eq. 9.20 that a PT for measuring AST shall have:

1. Similar surface emissivities
2. Similar form and size as the target specimen to have the same convection heat transfer coefficient
3. Well-insulated metal surfaces
4. Short response times

The first two items concern the heat transfers by radiation and convection, and the relation between the two. As described in Sect. 4.4 the AST depends on the radiation and convection heat transfer properties. Therefore the emissivity and the convection heat transfer coefficient should ideally be the same for the thermometer

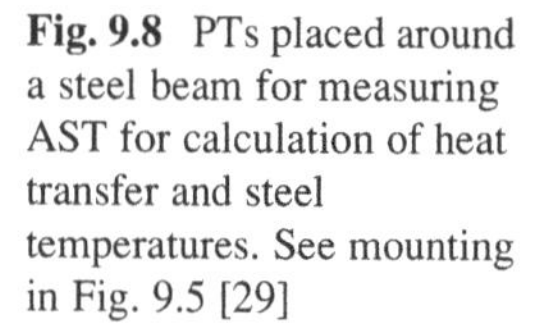

Fig. 9.8 PTs placed around a steel beam for measuring AST for calculation of heat transfer and steel temperatures. See mounting in Fig. 9.5 [29]

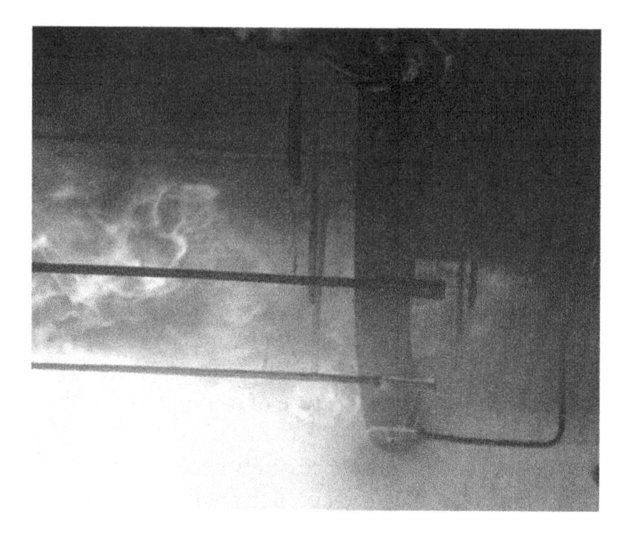

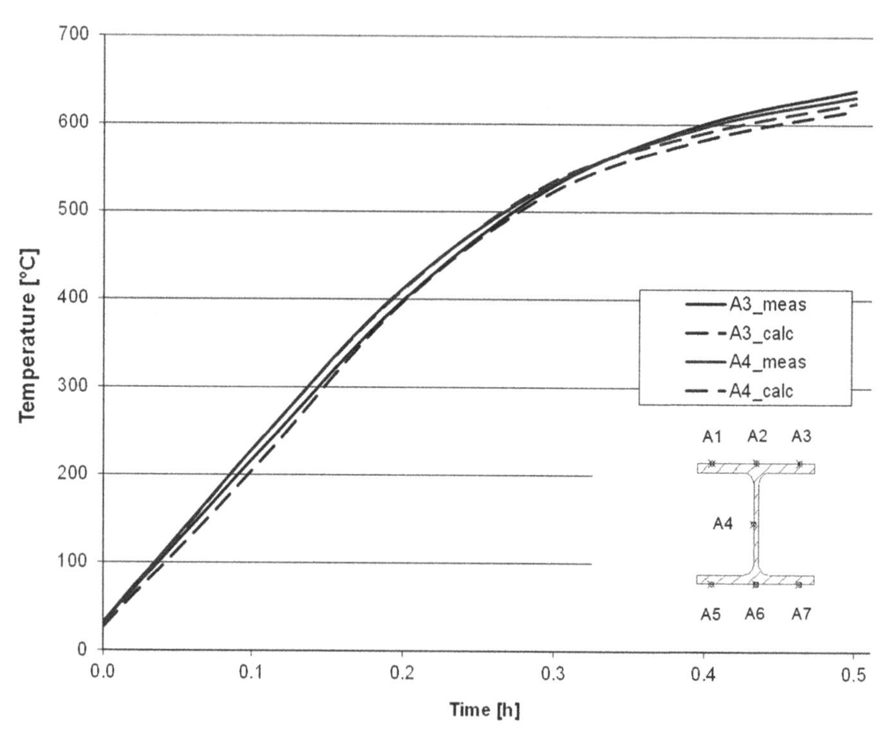

Fig. 9.9 Example of measured (*full line*) and calculated (*dashed line*) steel temperatures based on PT measurements as shown in Fig. 9.8

as for the target body of interest as far as it is practically possible. Thus for measuring the AST for calculating the heat transfer to, for example, a wall a standard PT might be a sensible compromise.

The third item, the PT should be well insulated or ideally perfectly insulated is of course not possible in practice. The reason for having a thin metal surface is to be able to measure the surface temperature accurately by fixing a thermocouple to the metal.

The fourth item is important for transient problems where the thermal exposure shall be followed as function of time.

The standard ISO 834 PT was designed for fire resistance furnace tests when being exposed to very high temperatures. It is therefore, on one hand, made very robust but not so well insulated as the temperature on the two sides of the PT does not differ very much in a furnace. However, when surrounded by air at ambient temperature and exposed to intense radiation at one side it must be better insulated for not losing heat from the exposed surface.

Figure 9.10 shows an example of two very well insulated PTs (so-called insPTs) designed to be used in ambient air. They measure thermal exposure and incident radiation in the vertical and horizontal directions, respectively. The plates which the thermocouples are fixed to are made of thin steel sheets (0.4 mm) to get quicker response times. To minimize the heat losses by conduction from the front to the back the insulation pads are thick and the sides of the steel plates have been partly cut out to avoid heat being conducted along the metal. On the back side this PT has a thicker more robust steel sheet for mounting purposes.

Figure 9.11 shows a comparison of incident heat flux measured with a Schmidt-Boelter heat flux meter and an insulated plate thermometer as shown in Fig. 9.10 applying Eq. 9.18 [34]. Note that the difference between the two methods of measuring incident heat flux is very small with the exception that the HFM responds much faster than the PT and therefore the measurement spikes which usually are not of interest to record.

Another alternative small PT has been developed for monitoring the thermal exposure in the ignition phase of a fire. This may be mounted flush at the surface of, for example, a combustible board as shown in Fig. 9.12. This so-called copper disc plate thermometer cdPT consists of a thin copper disc (Ø 12 mm and thickness 0.2 mm) backed with ceramic insulation mounted in an about 15 mm drilled hole.

To assure similar heat transfer properties the best way to design a PT may be to construct a 3-D dummy with a thin metal surface and filled with insulation. Figure 9.13 shows a pool fire and a steel cylinder simulating a piece of ammunition. Several thermocouples were fixed at various points of the dummy. It was placed in pool fire to record the thermal exposure of specimens placed in the flames.

Fig. 9.10 Two well-insulated plate thermometers (insPT) for measuring thermal exposures of horizontal and vertical surfaces, respectively, in air at ambient temperature

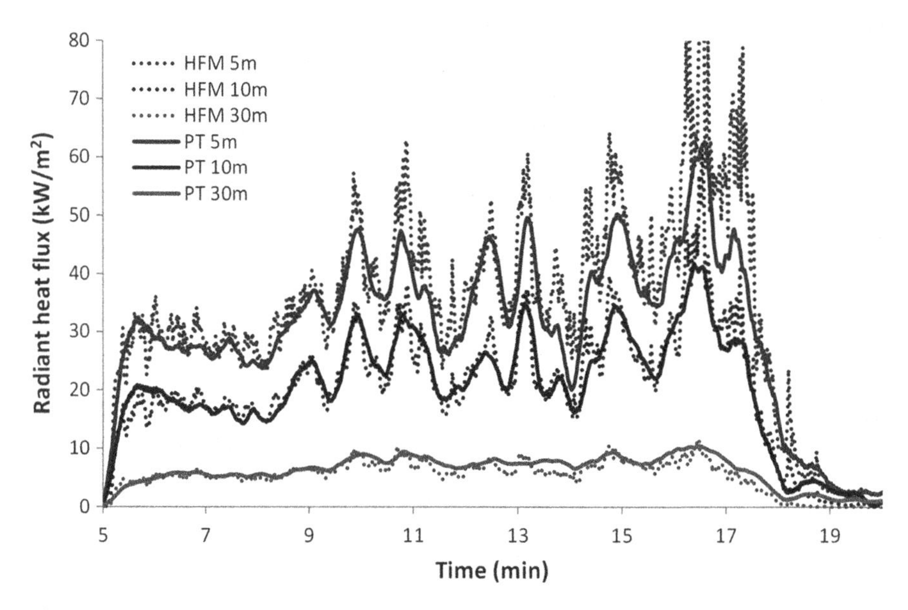

Fig. 9.11 Comparison of measurements at various distances of incident radiation with heat flux meters of Schmidt-Boelter type and with insPTs as shown in Fig. 9.10 applying Eq. 9.18

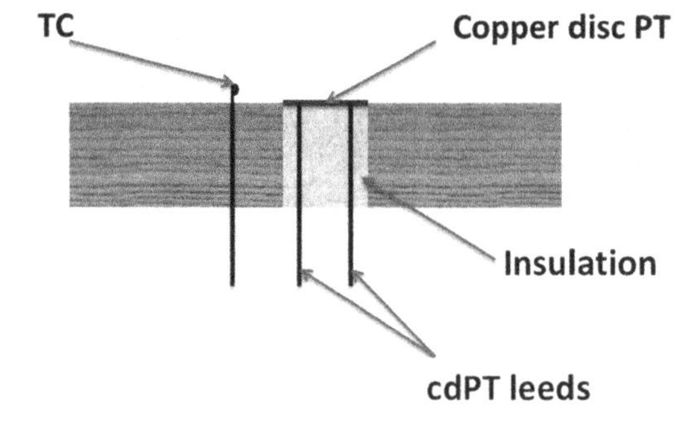

Fig. 9.12 Example of mounting of a copper disc plate thermometer (cdPT) flush at the surface of a board. Two thermocouple leads are welded to the copper disc. The back side of the disc is filled with insulation. A thin thermocouple (TC) is mounted nearby to measure gas temperature

Fig. 9.13 A steel cylinder dummy filled with insulation with thermocouples mounted on the surface were placed in a pool fire to register thermal exposure. (From a master thesis of Peter Möllerström and Björn Evers (2013), Luleå TU). (**a**) Pool fire (**b**) Steel cylinder dummy

Chapter 10
Post-Flashover Compartment Fires: One-Zone Models

FSE and design of structures and structural elements are in most cases made with a procedure including tests and classification systems. Fire resistance or endurance tests are specified in standards such as ISO 834, EN 1363-1 or ASTM E-119. In these standards time–temperature curves are specified representing fully developed compartment fires to be simulated in fire resistance furnaces for prescribed durations.

Alternatively design fires defined by their time–temperature curves may be obtained by making heat and mass balance analyses of fully developed compartment fires. Examples of that are given in the Eurocode 1 [36] where so-called *parametric fire curves* are defined. A number of significant simplifications and assumptions are then made to limit the number of input parameters and facilitate the calculations. Thus

1. The combustion rate is ventilation controlled, i.e. the heat release is proportional to the ventilation rate.
2. The fire compartment is ventilated by natural convection at a constant rate independent of temperature.
3. The gas temperature is uniform in the fire compartment.
4. The heat fluxes by radiation and convection to all surfaces of the compartment are equal and uniform.
5. The energy of the fuel is released entirely inside the compartment.
6. The fire duration is proportional to the amount of heat of combustion originally in the combustibles in the compartment, i.e. the fuel load.

All these assumptions are reasonable for a fully developed fire under ideal circumstances. The major parameters controlling the heat balance of fully developed compartment fires are then considered although they are violated more or less in real fires. Anyhow, by making certain parameter choices a set of time–temperature curves are obtained which in general yields design fires which are hotter and longer than could be anticipated in real fires or by more accurate numerical predictions.

U. Wickström, *Temperature Calculation in Fire Safety Engineering*,
DOI 10.1007/978-3-319-30172-3_10

The theory and assumptions outlined below follows the work of Thelandersson and Magnusson [36] and others but has been modified and reformulated according to later work by Wickström [37] made up for the basis for the parametric fire curves in Eurocode 1 [35]. See more on parametric fire curves in Sect. 12.2.

Below the fundamentals of the one-zone model theory are presented. The heat balance equation is then formulated in such a way that sometimes simple analytical solutions can be derived and in other cases general temperature calculation codes can be used to analyse compartments surrounded by boundaries of several layers and materials with properties varying with temperature as, for example, concrete containing water evaporating at 100 °C and having a thermal conductivity that decreases by 50 % during fire exposure.

10.1 Heat and Mass Balance Theory

The overall heat balance equation of a fully developed compartment fire as shown in Fig. 10.1 may be written as

$$\dot{q}_c = \dot{q}_l + \dot{q}_w + \dot{q}_r \tag{10.1}$$

where $\dot{q}_c$ is the heat release rate by combustion, $\dot{q}_l$ the heat loss rate by the flow of hot gases out of the compartment openings, $\dot{q}_w$ the losses to the fire compartment boundaries and $\dot{q}_r$ the heat radiation out through the openings. Other components of the heat balance equation are in general insignificant and not included in the approximate and simple analyses considered here.

When the temperature of the compartment rises, air and combustion products flow in and out of the compartment driven by buoyancy, i.e. the pressure difference Δp developed between the inside and outside of the compartment due to the gas temperature/density difference as indicated in Fig. 10.1. The mass of gases generated by the fuel when pyrolyzing is relatively small and therefore neglected. Hence the mass flow rate in $\dot{m}_i$ and out $\dot{m}_o$ of the compartment must be equal, denoted $\dot{m}_a$. Then by applying the Bernoulli theorem the flow rate of gases can be derived as approximately proportional to the opening area times the square root of its height for vertical openings.

$$\dot{m}_a = \alpha_1 A_o \sqrt{h_o} \tag{10.2}$$

where α_1 is a *flow rate coefficient*. A_o and h_o are the area and height of the openings of the compartment. The coefficient α_1 varies only slightly with the fire temperature over a wide range of temperatures relevant for fires and is therefore assumed constant [38]. In the presentation here only one vertical opening is assumed. For details on how multiple openings and horizontal openings can be considered see [35].

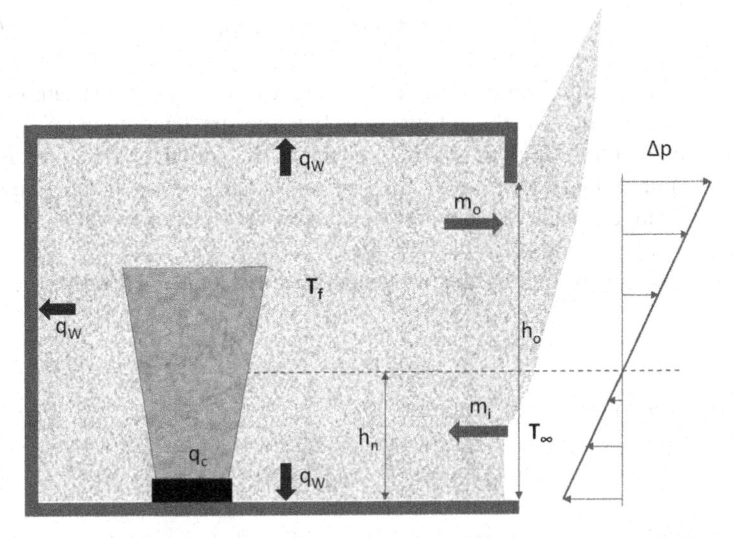

Fig. 10.1 One-zone model of a fully developed compartment fire with a uniform temperature T_f

As indicated in Fig. 10.1 hot fire gases are going out in the upper part of the opening and cool air is entering in the lower part. The level at which the direction of the flows are changing is called the neutral layer. As the outgoing flow of fire gases is hotter and less dense than the incoming air at ambient temperature, the neutral layer is below the middle of the opening, at about a third of the opening height. With the symbols shown in Fig. 10.1 that is $h_n \approx h_o/3$.

The combustion rate $\dot{q}_c$ inside the fire compartment is limited by the amount of air/oxygen available. Thus the fire is ventilation controlled and the combustion rate inside the compartment is proportional to the air flow, i.e.

$$\dot{q}_c = \chi\,\alpha_2\,\dot{m}_a = \chi\,\alpha_1\alpha_2 A_o\sqrt{h_o} \tag{10.3}$$

where the *combustion efficiency* χ is a reduction coefficient between zero and unity considering the burning efficiency, i.e. the fraction of the oxygen entering the compartment that is consumed by the combustion process inside the compartment. The *combustion yield* α_2 is the amount of energy released per unit mass of air in the combustion process. It is almost constant for combustible organic materials significant in fires with a value of about $13.2 \cdot 10^6$ W s/kg (per kg of oxygen). Then α_2 can be calculated assuming an oxygen content of 23 % in ambient air to be $3.01 \cdot 10^6$ W s/kg (per kg of air). (The fact that a constant amount of energy is released per unit weight of oxygen is also accounted for when measuring heat release rates by the so-called oxygen depletion technique, for example, in the cone calorimeter according to ISO 5660).

The first term on the right-hand side of Eq. 10.1 is the loss by flow of hot gas going out and being replaced by cooler gas. Hence $\dot{q}_l$ is proportional to the mass flow in and out of the compartment times the temperature rise of the fire, i.e.

$$\dot{q}_l = c_p\, \alpha_1 A_o \sqrt{h_o}\, \left(T_f - T_\infty\right) \tag{10.4}$$

where c_p is the specific heat capacity of the combustion gases at constant pressure (usually assumed equal to that of air) and T_f is the fire temperature. T_∞ is the ambient temperature which is assumed equal to the initial temperature T_i. The specific heat capacity of air c_p does not vary more than a few percentage over the temperature range considered and may be taken from textbooks such as [1, 2] at a temperature level of 800 °C to be $1.15 \cdot 10^3$ W s/(kg K).

For convenience of writing the fire temperature rise θ_f is introduced, i.e.

$$\theta_f = \left(T_f - T_i\right) \tag{10.5}$$

and the convection loss then becomes

$$\dot{q}_l = c_p\, \alpha_1\, A_o \sqrt{h_o}\, \theta_f \tag{10.6}$$

The second term on the right-hand side of Eq. 10.1, i.e. heat loss to the fire compartment boundaries $\dot{q}_w$ is assumed to be evenly distributed over the entire surrounding boundary area.

$$\dot{q}_w = A_t\, \dot{q}''_w \tag{10.7}$$

where A_t is the total enclosure area and $\dot{q}''$ the mean heat flux rate to the surrounding surfaces of the fire compartment. This term constitutes the inertia of the dynamic heat balance system as it changes with time depending on the temperature of the surrounding boundaries. It is significant in the beginning of a fire, and then it decreases when the temperature of the surrounding structure increases and gets closer to the fire temperature. For surrounding structures assumed to be thick it vanishes when thermal equilibrium is reached after long fire durations.

The third term on the right-hand side of Eq. 10.1, i.e. the heat loss by radiation directly out through the openings $\dot{q}_r$, may be calculated as

$$\dot{q}_r = \varepsilon_f\, A_o\, \sigma \left(T_f^4 - T_\infty^4\right) \tag{10.8}$$

where ε_f is the emissivity of the fire compartment at the opening here assumed to be one and therefore omitted below. This term is relatively small in the beginning of a fire when the fire temperature is moderate. It increases, however, by the forth power of the temperature and becomes considerable at later stages of fires when the temperature is high.

Now by inserting Eqs. 10.3, 10.4, 10.7 and 10.8 into Eq. 10.1 and after rearranging, the heat flux to the boundary surfaces becomes

$$\dot{q}''_w = c_p\, \alpha_1 O\left(\frac{\chi\alpha_2}{c_p} - \theta_f\right) + \frac{A_o}{A_t}\sigma\left(T_\infty^4 - T_f^4\right) \tag{10.9}$$

where O is named the *opening factor* defined as

$$O = \frac{A_o\sqrt{h_o}}{A_t} \tag{10.10}$$

The temperature of a ventilation controlled fire increases with time as the compartment boundary structures, ceiling, floor and walls heat up. If the compartment boundaries are assumed infinitely thick then when the compartment boundaries after a long time have been fully heated and steady-state thermal conditions can be assumed the heat losses to the boundary structure vanish. Notice in Eq. 10.9 that if the losses to the surrounding structure $\dot{q}_w$ and the radiation out the window $\dot{q}_r$ are negligible, the fire temperature depends only on the ratio between $\chi\alpha_2$ and c_p, i.e. ratio between product of the combustion efficiency and combustion yield, and the specific heat of the fire gases. It is, however, independent of the opening factor and the thermal properties of the surrounding structure. The parameter is here named the *ultimate fire temperature* θ_{ult}:

$$\theta_{ult} = \frac{\chi\alpha_2}{c_p} \tag{10.11}$$

The values of all the parameters introduced above vary only slightly with temperature and are therefore here assumed to remain constant. Commonly assumed values are summarized in Table 10.1.

In Eq. 10.9 the parameter groups $(c_p\alpha_1 O)$, $(\chi\alpha_2/c_p)$ and (A_o/A_t) are constants, and (T_∞) is a known boundary temperature. Therefore this equation is analogous to a boundary condition of the third kind as outlined in Sect. 3.2.3. If the radiation directly out through the openings is neglected (second term on the right-hand side of Eq. 10.9) even analytical solutions can sometimes be obtained as shown in Sect. 10.2.

Alternatively Eq. 10.9 may be written as

$$\dot{q}''_w = h_{f,c}\left(T_{ult} - T_f\right) + h_{f,r}\left(T_\infty - T_f\right) \tag{10.12}$$

or as

$$\dot{q}''_w = h_{f,c}\left(\theta_{ult} - \theta_f\right) + h_{f,r}\theta_f \tag{10.13}$$

Here $h_{f,c}$ is named *fire compartment convection heat transfer coefficient*, identified as

Table 10.1 Values of physical parameters and parameter groups

Name	Notation	Value	Units
Flow rate coefficient	α_1	0.5	$kg/(s\ m^{5/2})$
Combustion yield coefficient	α_2	$3.01 \cdot 10^6$	W s/kg
Specific heat capacity of air	c_p	1150	W s/(kg K)
Combustion efficiency	χ		–
Ultimate fire temperature increase	$\theta_{ult} = \frac{\chi\alpha_2}{c_p}$	1325 ($\chi = 0.506$)	K
Fire convective heat transfer coefficient	$h_{f,c} = c_p\ \alpha_1 O$	575·O	$W/(m^2\ K)$

$$h_{f,c} = c_p\ \alpha_1 O \tag{10.14}$$

and *fire compartment radiation heat transfer coefficient* $h_{f,r}$ is identified as

$$h_{f,r} = \frac{A_o}{A_t}\sigma\left(T_\infty^2 + T_f^2\right) \cdot \left(T_\infty + T_f\right) \tag{10.15}$$

The corresponding *fire compartment thermal resistances* are defined as

$$R_{f,c} = \frac{1}{h_{f,c}} = \frac{1}{c_p\ \alpha_1 O} \tag{10.16}$$

and

$$R_{f,r} = \frac{1}{h_{f,r}} = \frac{1}{\frac{A_o}{A_t}\sigma\left(T_\infty^2 + T_f^2\right) \cdot \left(T_\infty + T_f\right)} \tag{10.17}$$

The ultimate fire temperature θ_{ult} will generally not appear in reality. It is introduced to facilitate the development and explanation of the compartment fire models.

The heat transfer to the surrounding structure expressed in terms of the fire temperature may be written as

$$\dot{q}''_w = h_{i,c}\left(T_f - T_s\right) + h_{i,r}\left(T_f - T_s\right) \tag{10.18}$$

where $h_{i,c}$ is the convection heat transfer coefficient and $h_{i,r}$ the radiation heat transfer coefficient between the fire gases and the compartment boundary. The latter is defined as

$$h_{i,r} = \varepsilon_s\sigma\left(T_f^2 + T_s^2\right) \cdot \left(T_f + T_s\right) \tag{10.19}$$

where ε_s is the emissivity of the fire compartment inner surface. Then the combination of Eqs. 10.12 and 10.18 can be illustrated by an electric circuit analogy as shown in Fig. 10.2 where the resistances $R_{f,c}$ and $R_{f,r}$ are defined in Eqs. 10.16 and

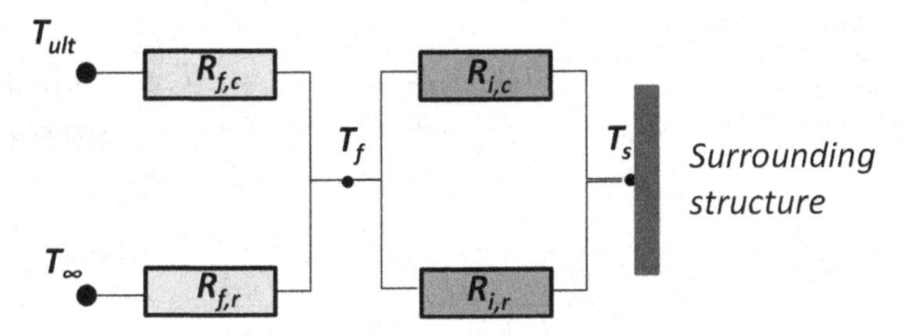

Fig. 10.2 Electric circuit analogy model of a fire compartment boundary

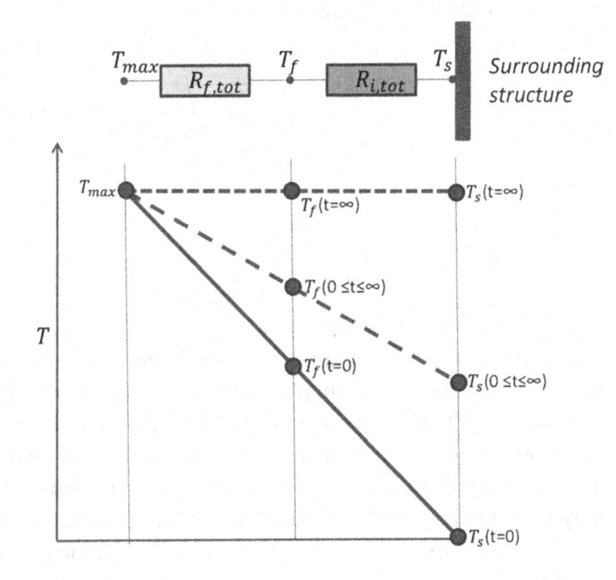

Fig. 10.3 Electric circuit analogy model of a fire compartment boundary with two heat transfer resistances in series

10.17, and $R_{i,c}$ and $R_{i,r}$ are the inverses of the corresponding heat transfer coefficients as defined by Eqs. 10.18 and 10.19.

The two temperatures T_{ult} and T_∞ may be reduced to one resultant temperature T_{max} which is a weighted mean value of the two. Compare with adiabatic surface temperature of Sect. 4.4. Then the electric circuit of Fig. 10.2 can be reduced that of Fig. 10.3. The analogy between heat transfer and electric circuit parameters is described in Sect. 1.2 where also rules for combining resistances in series and parallel are given.

T_{max} is the maximum temperature a compartment fire can reach when the losses to the boundaries vanish. It can be calculated by putting $\dot{q}''_w = 0$ in Eq. 10.12 and solving for θ_f.

As there is no thermal heat capacity involved, the heat flux may now be written in two ways as

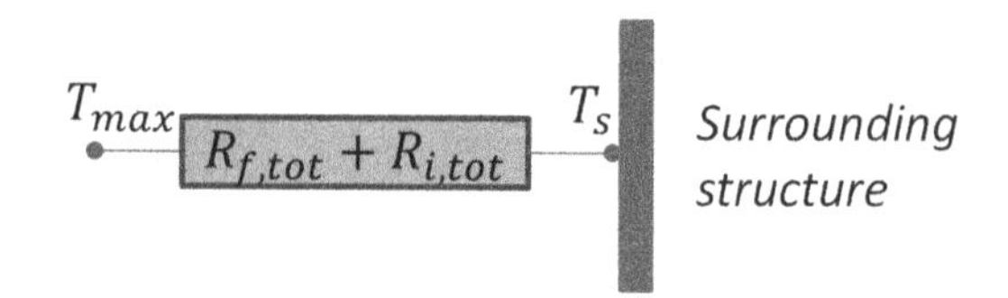

Fig. 10.4 Electric circuit analogy model of a fire compartment boundary with two heat transfer resistances in series

$$\dot{q}''_w = \frac{(T_{max} - T_f)}{R_{f,tot}} = \frac{(T_{max} - T_{i,s})}{R_{f,tot} + R_{i,tot}} \tag{10.20}$$

Observe that the radiation heat transfer coefficient must be calculated at the absolute temperatures T_{max} and T_f. Thus

$$R_{f,tot} = \frac{1}{c_p\, \alpha_1 O + \frac{A_o}{A_t}\sigma\left(T_{max}^2 + T_f^2\right) \cdot (T_{max} + T_f)} \tag{10.21}$$

and then

$$R_{i,tot} = \frac{1}{h_{i,c} + h_{i,r}} = \frac{1}{h_{i,c} + \varepsilon_s\sigma\left(T_f^2 + T_{i,s}^2\right) \cdot (T_f + T_{i,s})} \tag{10.22}$$

According to Eq. 10.20 the two thermal boundary resistances in series as shown in Fig. 10.3 can then be summarized into one as shown in Fig. 10.4.

Equation 10.20 is a third kind of boundary condition (see Sect. 1.1.3) with the heat transfer coefficient equal to the reciprocal of the heat transfer resistance. With this boundary condition combined with a thermal model of the boundary structure may its temperature be calculated including its surface temperature.

The fire temperature T_f can thereafter be obtained as the weighted mean temperature of T_{max} and T_s as

$$T_f = \theta_f + T_i = \frac{T_s\, R_{f,tot} + T_{max}\, R_{i,tot}}{R_{f,tot} + R_{i,tot}} \tag{10.23}$$

If the thermal resistances $R_{f,tot}$ and $R_{i,tot}$ may be assumed constant, analytical solutions for T_s can be derived for surrounding structures being semi-infinitely thick or having its heat capacity lumped in a core as is shown below.

The highest fire temperature that can be reached in a fire compartment occurs when surrounding structures are fully heated and do not absorb any more heat, i.e. $\dot{q}''_w$ vanishes. Then the fire temperature and the surface temperature becomes equal to T_{max}. If in turn the radiation directly out through the openings can be neglected as well, the maximum fire temperature becomes T_{ult}.

An observation is that according to this theory an instant fire temperature rise occurs when the fire begins. Then the fire temperature immediately T_f^0 rises to

$$T_f^0 = \frac{R_{i,tot}}{R_{f,c} + R_{i,tot}} T_{max} \tag{10.24}$$

This immediate temperature rise is of course physically unlikely for the very initial phase as a heat release yielding flashover cannot start suddenly in reality but after some time has elapsed the approximate predictions as given by the above theory applies.

10.2 Solution of the Fire Compartment Temperature

The boundary condition as defined above includes the two heat transfer resistances in a series, one artificial and one physical as indicated by Fig. 10.4. To solve for the surface temperature and then calculate the fire temperature according to Eq. 10.23, a thermal model of the compartment boundary structure is needed. The surface temperatures may then be calculated with various methods depending on whether the model parameters may vary with temperature. When either heat transfer coefficients or material properties vary with temperature, the problem becomes non-linear and then numerical tools such as finite element programs need to be used. Boundary structures of several layers of different materials etc. may then also be considered. Spreadsheet calculations using programs such as MS-Excel are very useful when analysing fire compartments with boundaries where lumped heat can be assumed.

Numerically exact analytical expressions can be derived for two types of boundary constructions being considered in the next sections, namely, structures assumed either semi-infinitely thick or having a core where the thermal mass is concentrated (lumped heat). Then the elementary procedures presented in Sects. 3.2 and 3.1, respectively, may be applied.

In Table 10.1 a summary of values of physical parameters and parameter groups are given. These are used throughout the presentation below.

10.2.1 *Semi-infinitely Thick Compartment Boundaries*

Fire compartment boundaries are in most cases assumed thermally thick. The heat transferred to the surfaces are then stored in the surrounding structures, and the effects of heat lost on the outside of the structure is neglected.

As indicated by Fig. 10.4 may the boundary condition be expressed by two thermal boundary resistances in series which can be added up and a complete thermal model becomes as indicated by Fig. 10.5.

This is a semi-infinite body with a third kind of boundary condition. To compute the surface temperature generally numerical temperature calculation methods are

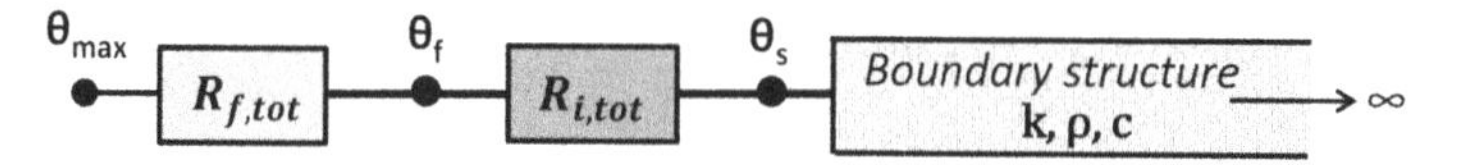

Fig. 10.5 Electric circuit analogy model of a fire compartment with infinitely thick walls

needed such as finite element methods. The fire temperature is then calculated as the weighted mean temperature of T_{max} and T_s according to Eq. 10.23.

However, if the following assumptions are made, the problem becomes linear and analytical solutions can be derived:

1. The heat radiated directly out the openings, $\dot{q}_r$, is neglected or is directly proportional to the difference between the fire temperature T_f and the ambient temperature T_∞, i.e. $h_{f,tot}$ and its reciprocal $R_{f,tot}$ are constant.
2. The heat transfer by radiation and convection to the surrounding boundaries is assumed proportional to the difference between the fire T_f and boundary surface temperatures T_s, i.e. $h_{i,tot}$ and its reciprocal $R_{i,tot}$ are constant.

The surface temperature rise can be calculated according to Eq. 3.35 in Sect. 3.2.3. Thus

$$\theta_s = \theta_{max} \cdot \left[1 - e^{\frac{t}{\tau_f}} \cdot erfc\left(\sqrt{\frac{t}{\tau_f}}\right)\right] \tag{10.25}$$

where the parameter τ_f may be identified as a *fire compartment time constant* for infinitely thick walls in analogy with Eq. 3.34.

$$\tau_f = \frac{k \cdot \rho \cdot c}{\left(\frac{1}{R_{f,tot} + R_{i,tot}}\right)^2} = k \cdot \rho \cdot c \cdot \left(R_{f,tot} + R_{i,tot}\right)^2 \tag{10.26}$$

as the reciprocal of the heat transfer resistance by definition is equal to the heat transfer coefficient. $R_{f,tot}$ depends on $R_{f,c}$ and $R_{f,r}$. The former is always constant according to Eq. 10.16, and by assuming constant fire temperature $T_f = T_f^\vee$ in Eq. 10.17, a constant $R_{f,r}$ could be calculated as well. Too high assumed $T_f^\vee$-values will yield overestimated the heat losses by radiation out the openings and therefore underestimated fire temperatures, and vice versa. θ_{max} is the temperature rise which is obtained when the wall are fully heated and no heat is transferred to boundary surfaces. Given constant values of $R_{f,c}$ and $R_{f,r}$, θ_{max} is constant and can be calculated as

$$\theta_{max} = \frac{\theta_{ult}\, R_{f,r}}{R_{f,r} + R_{i,c}} = \frac{\theta_{ult}}{1 + \frac{R_{f,c}}{R_{i,r}}}. \tag{10.27}$$

By inserting the opening factor from Eq. 10.10 and assuming parameter values according to Table 10.1 a

$$\frac{R_{f,c}}{R_{f,r}} = \frac{\sigma \cdot \left(T_{\infty}^{2} + \overset{\vee}{T}_{f}^{2}\right) \cdot \left(T_{\infty} + \overset{\vee}{T}_{f}\right)}{575\sqrt{h_o}} \tag{10.28}$$

where $\overset{\vee}{T}$ is the assumed temperature level. Then by inserting Eq. 10.28 into Eq. 10.27

$$\theta_{max} = \frac{\theta_{ult}}{1 + \frac{\sigma \cdot \left(T_{\infty}^{2} + \overset{\vee}{T}_{f}^{2}\right) \cdot \left(T_{\infty} + \overset{\vee}{T}_{f}\right)}{575\sqrt{h_o}}}. \tag{10.29}$$

Notice that θ_{max} increases with the square root of the opening height h_o but is independent of A_o and A_t.

The fire temperature rise vs. time may be obtained as the weighted average of θ_{max} and θ_s in analogy with Eq. 10.23 as

$$\theta_f = \frac{\theta_s R_{f,tot} + \theta_{max} R_{i,tot}}{R_{f,tot} + R_{i,tot}}. \tag{10.30}$$

Now by inserting Eq. 10.25 into Eq. 10.26, the fire temperature development becomes

$$\theta_f = \frac{\theta_{ult}}{1 + \frac{R_{i,tot}}{R_{f,tot}}} \left\{ \left[1 - e^{\frac{t}{\tau_f}} \cdot erfc\left(\sqrt{\frac{t}{\tau_f}}\right) \right] + \frac{R_{i,tot}}{R_{f,tot}} \right\}. \tag{10.31}$$

Constant values of $R_{i,tot}$ and $R_{f,tot}$ may be obtained from Eqs. 10.21 and 10.22, respectively, for a given fire temperature $T_f = \overset{\vee}{T}_f$. Then by inserting the ratio $\frac{R_{i,tot}}{R_{f,tot}}$ into Eq. 10.31, a very handy closed form solution of the fire temperature development vs. time is obtained.

An interesting observation is that the standard design time–temperature curves may be derived by prescribing a maximum temperature rise $\theta_{ult} = \chi\alpha_2/c_p = 1325\,°C$ and a fire compartment time constant $\tau_f = 1200\,s$. This time constant may be calculated based on quite reasonably assumed input parameters for surrounding boundary properties grouped into the thermal inertia $(k \cdot \rho \cdot c)$, the opening factor (O) and the heat transfer resistance between the fire gases and the surrounding boundaries $(R_{i,tot})$. Then Eq. 10.25 yields the fire temperature rise as

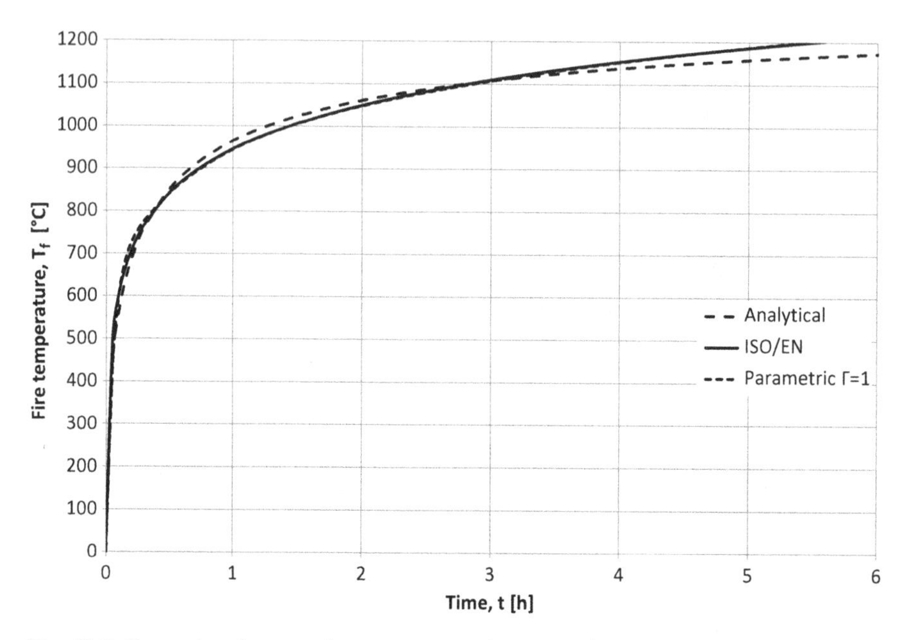

Fig. 10.6 Comparison between fire temperature rises according to the analytical expression (Eq. 10.32), the standard ISO 834 curve and the parametric fire curve for $\Gamma = 1$ according to EN 1991-1-2

$$\theta_f = 1325 \left[1 - e^{\left(\frac{t}{1200}\right)} \cdot erfc\sqrt{\frac{t}{1200}} \right]. \tag{10.32}$$

which is very close to the EN 1363-1 and ISO 834 standard curves as well as the heating phase of the parametric fire curve according to EN 1991-1-2 for the compartment factor $\Gamma = 1$, see Sect. 12.2, as shown in Fig. 10.6.

Example 10.1 Calculate the fully developed fire temperature rise after 60 min in a compartment surrounded by concrete. Assume an ultimate temperature $\theta_{ult} = 1325\,°\mathrm{C}$ and an opening factor of $O = 0.04\ \mathrm{m}^{½}$, material properties according to Table 1.2 and other physical parameters according to Table 10.1.

(a) Neglecting the effects of heat transfer resistance between the fire gases and the surrounding boundaries and the radiation directly out the window.
(b) Neglecting the effects of radiation directly out the window but not the effects of heat transfer resistance between the fire gases and the surrounding boundaries. Assume a total heat transfer coefficient $h_{tot,i} = 200\,\mathrm{W/(m^2\,K)}$.
(c) Considering both the effects of heat transfer resistance between the fire gases and the surrounding boundaries and the radiation directly out the window. The ratio between the opening area and the total surrounding area $A_o/A_t = 0.06$. Assume a $h_{f,r} = 70\,\mathrm{W/(m^2\,K)}$.

Solution

(a) ($R_{i,tot} = 0$ and $R_{f,r} = 0$). Equation 10.16 yields $R_{f,c} = R_{f,tot} = \frac{1}{575 \cdot 0.04} = 0.0435\,(\mathrm{m^2 K})/\mathrm{W}$ and Eq. 10.26 yields $\tau_f = \frac{3530000}{(575 \cdot 0.04)^2} = 6673\,\mathrm{s}$. Thus $t/\tau_f = \frac{3600}{6673} = 0.54$ and Eq. 10.25 (and Table 3.3) yields $\theta_f = \theta_s = 0.49 \cdot 1325 = 649\,°\mathrm{C}$.

(b) ($R_{f,r} = 0$). $R_{i,tot} = \frac{1}{200} = 0.005\,(\mathrm{m^2 K})/\mathrm{W}$ Eq. 10.26 yields $\tau_f = 3530000 \cdot (0.0435 + 0.005)^2 = 8303\,\mathrm{s}$ and $t/\tau_f = \frac{3600}{8303} = 0.433$ and Eq. 10.25 yields the surface temperature $\theta_s = 1325 \cdot 0.46 = 610\,°\mathrm{C}$. Then the fire temperature can be obtained from Eq. 10.30 as $\theta_f = \frac{610 \cdot 0.0435 + 1325 \cdot 0.005}{0.0435 + 0.005} = \frac{42.55}{0.0578} = 683\ °\mathrm{C}$.

(c) According to Eq. 10.15 $h_{f,r} = \frac{A_o}{A_{tot}} \sigma \left(T_\infty^2 + T_f^2\right) \cdot (T_\infty + T_f) \approx 0.03 \cdot 70 = 2.1\ \mathrm{W}/(\mathrm{m^2 K})$ and according to Table 10.1 $h_{f,c} = \frac{1}{R_{f,c}} = 23\,\mathrm{W}/(\mathrm{m^2 K})$ and thus $R_{f,tot} = \frac{1}{2.1 + 23} = 0.040\,(\mathrm{m^2 K})/\mathrm{W}$. Then Eq. 10.27 yields $\theta_{max} = \frac{23 \cdot 1325}{23 + 2.1} = 1214\,°\mathrm{C}$ and Eq. 10.26 yields $\tau_f = k \cdot \rho \cdot c \cdot \left(R_{f,tot} + R_{i,tot}\right)^2 = 3530000 \cdot [0.040 + 0.005]^2 = 7148$ s and $t/\tau_f = \frac{3600}{6227} = 0.504$ and Eq. 10.25 yields the surface temperature $\theta_s = 0.48 \cdot 1214 = 583\ °\mathrm{C}$. Then the fire temperature can be obtained from Eq. 10.30 as $\theta_f = \frac{\theta_s\, R_{f,tot} + \theta_{max}\, R_{i,tot}}{R_{f,tot} + R_{i,tot}} = \frac{583 \cdot 0.040 + 1214 \cdot 0.005}{0.040 + 0.005} = 652\ °\mathrm{C}$.

Comment: Notice that the various fire temperatures are obtained depending on the levels of completeness of the calculation model.

10.2.2 Insulated and Uninsulated Boundaries with a Metal Core

Analytical solutions of the fire temperatures may also be obtained when the fire compartment is assumed surrounded by structures consisting a metal core where the all the heat capacity is concentrated. Then the heat capacity per unit area C_{core} is lumped into the core as indicated in Fig. 10.7. The heat capacity of any insulating material is either neglected or assumed included in the heat capacity of the core.

Figure 10.8 shows an electric circuit analogy model of how the fire, the core and the inner and outer surface temperatures can be calculated. As all inertia is lumped into the core, the heat flux is constant on either side of the core due to the requirement of heat flux continuity. Hence the temperature differences between various positions are proportional to the corresponding thermal resistances. The three graphs of the figure indicate the temperature rises initially, after some finite time and after a very long time, respectively. Notice that according to the theory the fire and the inner fire exposed surface temperatures increase instantaneously at $t = 0$ according to Eq. 10.29.

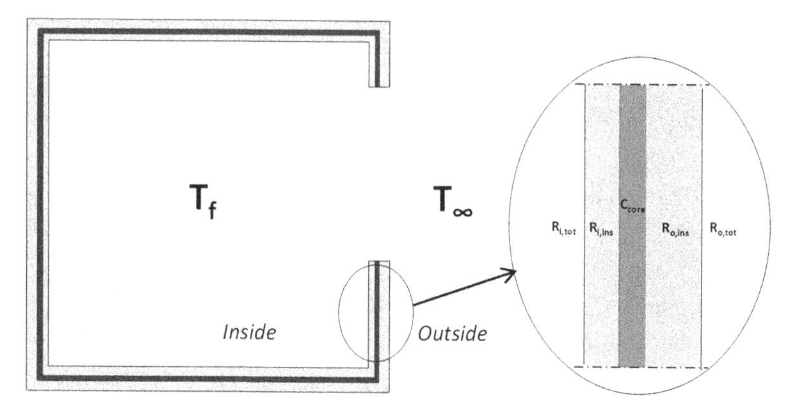

Fig. 10.7 A fire compartment surrounded by a structure with its heat capacity C_{core} assumed concentrated/lumped to a metal core. Thermal resistances of insulation materials R_i and R_o are assumed on the fire inside and outside, respectively

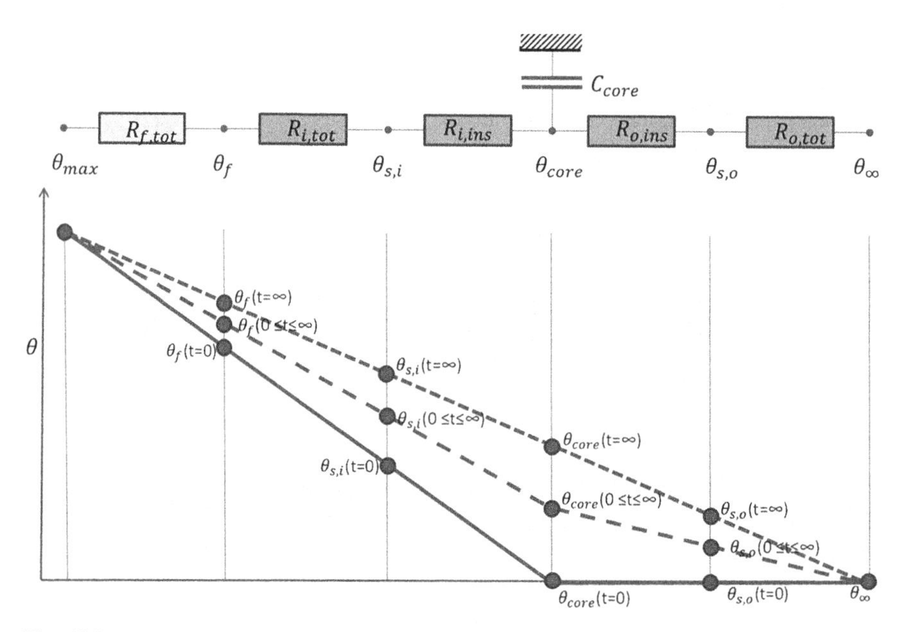

Fig. 10.8 Electric circuit analogy of fire compartment model with a thin surrounding structure assuming lumped heat capacities. Relative temperatures at various points initially (t = 0), after some time ($0 \leq t \leq \infty$) and after a very long time ($t = \infty$) are indicated

The parameters of Fig. 10.8 are summarized in Table 10.2. The maximum compartment fire temperature considering radiation out through openings θ_{max} can be obtained from Eq. 10.12 or Eq. 10.23, the fire compartment heat transfer resistance $R_{f,tot}$ from Eq. 10.21 and the heat transfer resistance between fire and

Table 10.2 Summary of the parameters of Fig. 10.8

Notation	Parameter	Definition
θ_{max}	Maximum temperature—no losses to surfaces	Eq. 10.12, Eq. 10.23 or Eq. 10.29
$R_{f,tot}$	Fire heat transfer resistance	Eq. 10.21
$R_{i,tot}$	Transfer resistance, fire—inside surface	Eq. 10.22
$R_{i,ins}$	Resistance inside insulation	Eq. 10.33
$R_{o,ins}$	Resistance outside insulation	Eq. 10.34
$R_{o,tot}$	Transfer resistance, outside surface—surroundings	Eq. 10.35
C_{core}	Heat capacity of the core	Eq. 10.36

surrounding surfaces $R_{i,tot}$ from Eq. 10.22. The insulation resistances at the inside and outsider of the core $R_{i,ins}$ and $R_{o,ins}$ can be calculated as

$$R_{i,ins} = \frac{d_{i,ins}}{k_{i,ins}} \tag{10.33}$$

and

$$R_{o,ins} = \frac{d_{o,ins}}{k_{o,ins}} \tag{10.34}$$

where $d_{i,ins}$ and $d_{o,ins}$ are the thicknesses of the inside and outside insulations, respectively, and $k_{i,ins}$ and $k_{o,ins}$ are the corresponding conductivities. The outside heat transfer resistance can be calculated as

$$R_{o,tot} = \frac{1}{h_{o,c} + \varepsilon_{o,s}\sigma\left(T_{o,s}^2 + T_\infty^2\right) \cdot \left(T_{o,s} + T_\infty\right)} \tag{10.35}$$

where $h_{o,c}$ is the convection heat transfer coefficient at the outside surface and $\varepsilon_{o,s}$ the emissivity of the outside surface. The conduction resistance of the metal core is neglected. C_{core} is the heat capacity per unit area of the core, i.e.

$$C_{core} = d_{core} c_{core} \rho_{core}. \tag{10.36}$$

where d_{core}, c_{core} and ρ_{core} are the thickness, specific heat and density of the core, respectively

The dynamic heat balance of the core can now be written as

$$C_{core}\frac{d\theta_{core}}{dt} = \left(R_{f,tot} + R_{i,tot} + R_{i,ins}\right)\left(\theta_{max} - \theta_{core}\right) - \left(R_{o,ins} + R_{o,tot}\right) \cdot \left(\theta_{core} - \theta_{\infty}\right) \quad (10.37)$$

and the core temperature can be numerically solved by the forward difference recursion formula

$$\theta_{core}^{i+1} = \theta_{core}^{i} + \frac{\Delta t}{C_{core}}\left[\left(R_{f,tot} + R_{i,tot} + R_{i,ins}\right)\left(\theta_{max} - \theta_{core}\right) - \left(R_{o,ins} + R_{o,tot}\right) \cdot \left(\theta_{core} - \theta_{\infty}\right)\right] \quad (10.38)$$

where Δt is a chosen time increment. This recursion formula can be coded in spreadsheet programs such as MS-Excel. The temperature-dependent parameter may be updated along with the calculations.

If all the parameters are assumed constant, then the temperature development can be calculated analytically. Thus, the core temperature rise may be obtained as a function of time as (see Sect. 3.1.2)

$$\theta_{core} = \theta_{max}\left[\frac{R_{o,tot} + R_{o,ins}}{R_{f,tot} + R_{i,tot} + R_{i,ins} + R_{o,tot} + R_{o,ins}}\right]\left(1 - e^{-t/\tau_f}\right) \quad (10.39)$$

where the fire compartment time constant τ_f is

$$\tau_f = \frac{C_{core}}{\frac{1}{R_{f,tot}+R_{i,tot}+R_{i,ins}} + \frac{1}{R_{o,tot}+R_{o,ins}}} \quad (10.40)$$

When the core temperature has been calculated, the fire temperature may be calculated. As the heat capacity of the insulation is assumed to be negligible, the compartment fire temperature rise can be calculated as a weighted average between θ_{max} and θ_{core} (see Fig. 10.8). Thus

$$\theta_f = \frac{\left(R_{i,tot} + R_{i,ins}\right)\theta_{max} + R_{f,tot}\,\theta_{core}}{R_{f,tot} + R_{o,tot} + R_{o,ins}} \quad (10.41)$$

and after inserting θ_{core} according to Eq. 10.39 and rearranging

$$\theta_f = \frac{\theta_{max}}{1 + \frac{R_{i,tot} + R_{i,ins}}{R_{f,tot}}}\left[\frac{R_{o,tot} + R_{o,ins}}{R_{f,tot} + R_{i,tot} + R_{i,ins} + R_{o,tot} + R_{o,ins}}\left(1 - e^{-t/\tau_f}\right) + \frac{R_{i,tot} + R_{i,ins}}{R_{f,tot}}\right] \quad (10.42)$$

Notice that this expression is similar to the corresponding Eq. 10.31 for semi-infinite boundaries.

In a corresponding way according to the law of proportion may the fire exposed surface temperature be calculated as

$$\theta_{s,i} = \frac{R_{i,ins}\,\theta_{max} + \left(R_{f,tot} + R_{i,tot}\right)\theta_{core}}{R_{f,c} + R_{i,tot} + R_{i,ins}} \tag{10.43}$$

and non-exposed surface temperature as

$$\theta_{s,o} = \frac{R_{o,tot}\,\theta_{core}}{R_{o,tot} + R_{o,ins}} \tag{10.44}$$

The maximum fire temperature that can be reached asymptotically after long fire durations depends on the insulation of the compartment and the heat transfer resistances. It can be calculated as

$$\theta_f^{max} = \frac{R_{i,tot} + R_{i,ins} + R_{o,tot} + R_{o,ins}}{R_{f,tot} + R_{i,tot} + R_{i,ins} + R_{o,tot} + R_{o,ins}}\,\theta_{max} \tag{10.45}$$

The fire development is generally very fast for thin structures and the maximum is reached quickly.

Example 10.2 Estimate the maximum post-flashover fire temperature T_f^{max} of an uninsulated steel container. It has a steel thickness $d_{core} = 3$ mm, an opening height and area $h_o = 2.5$ m and $A_o = 5$ m^2, and a total area $A_t = 125$ m^2. Assume $\theta_{ult} = 1325$ °C, the internal and external heat transfer coefficients due to radiation and convection $h_{i,tot} = 1/R_{i,tot} = 100$ W/(m^{2}K) and $h_{o,tot} = 1/R_{o,tot} = 25$ W/(m^{2}K). Estimate the fire temperature development vs. time.

Solution
Assume a maximum fire temperature $T_f^{max} = 900\,°\mathrm{C} = 1173\,\mathrm{K}$ for the estimate of $R_{f,tot}$. Then Eq. 10.29 yields $\theta_{max} = \frac{1325}{1+\frac{\sigma\cdot(293^2+1173^2)\cdot(293+1173)}{575\ \sqrt{2.5}}} = 1169$ °C. Then Eq. 10.21 and Table 10.1 yield $R_{f,tot} = \frac{1}{575\cdot\frac{5\sqrt{2.5}}{125}+\frac{5}{125}\sigma\left[(1169+273)^2+973^2\right]\cdot(1169+273+973)} = 2\ \mathrm{m^2K/W}$.
Then Eq. 10.36 yields $\theta_f^{max} = \frac{1/100+0+1/25+0}{0.022+1/100+1/25}\,1169 = 809$ °C. A new estimate $T_f^{max} = 850$ °C $= 1123$ K yields $\theta_{max} = 1184$ °C, $R_{f,tot} = 0.0178$ and $\theta_f^{max} = 0.731184 = 873$ °C.

Comment: Thus the temperature rise in an uninsulated container as described will never exceed about 870 °C (θ_f^{max}). For a well-insulated container with the same geometry the fire temperature rise may reach about 1180 °C (θ_{max}).

The temperature development can be estimated by applying Eq. 10.42. The time constant $\tau_f = \frac{0.003\cdot 460\cdot 7850}{1/(0.0178+0.01)+25} = 177$ s and then if $R_{f,tot}$ is assumed constant as calculated above $\theta_f = \frac{1184}{1+\frac{0.01}{0.0178}}\left[\frac{0.04}{0.0178+0.01+0.04}\left(1-e^{-t/177}\right)+\frac{0.01}{0.0178}\right] = 447\cdot\left(1-e^{-t/177}\right)+426$ °C.

Comment: This solution shows that the time constant is very short, less than 3 min, and that very beginning of the fire ($t = 0$) the calculated fire temperature rise is significant, $\theta_f = 426$ °C, and that after a long time ($t \to \infty$) $\theta_f \to 873$ °C as stated above (θ_f^{max}).

Example 10.3 A fire compartment is surrounded by a 3-mm-thick steel sheet structure with a 12-mm-thick gypsum board mounted on both sides of the core. The opening is factor O = 0.08 $m^{1/2}$. The heat transfer coefficient at the fire exposed and the unexposed sides are assumed to be constant, i.e. $h_{i,tot} = 200\,W/(m^2\,K)$ and $h_{o,tot} = 40\,W/(m^2\,K)$, respectively. Neglect the radiation directly out the opening, i.e. $R_{f,tot} = R_{f,c}$.

(a) Calculate the ultimate fire temperature rise θ_{ult} assuming a combustion efficiency of 50 %.
(b) Calculate the maximum fire θ_f^{max} and core θ_{core}^{max} temperature rises.
(c) Calculate the fire temperature θ_f and the inner fire exposed surface temperature $\theta_{s,i}$ at time t = 0 according to the model.
(d) Calculate the core θ_{core} and fire θ_f temperature rises after 300 s of flashover.
(e) Plot as a function of time of the temperature rises, ultimate θ_{ult}, fire θ_f, inner surface $\theta_{s,i}$, core θ_{core}, outer surface $\theta_{s,o}$.

Use parameter values as given in Tables 1.2 and 10.1.

Solution

(a) Equation 10.11 yields $\theta_{ult} = 1309$ °C.
(b) The thermal resistances over a unit area $(R_{i,tot} + R_{i,ins}) = 1/200 + 0.012/0.5 = 0.029\ (m^2\,K)/W$ and $(R_{o,tot} + R_{o,ins}) = \frac{1}{40} + 0.012/0.5 = 0.049\ (m^2K)/W$. $R_{f,c} = R_{f,tot} = 1.74 \cdot \frac{10^{-3}}{0.08} = 0.022 (m^2\,K)/W$. Then Eq. 10.45 yields $\theta_f^{max} = \frac{0.029+0.049}{0.022+0.029+0.049} \cdot 1309 = 1021$°C and Eq. 10.39 yields $\theta_{core}^{max} = \frac{0.049}{0.022+0.049+0.029} \cdot 1309 = 641$ °C.
(c) Equation 10.24 yields $\theta_f^0 = \frac{0.029}{0.022+0.029} \cdot 1309 = 744$ °C, and $\theta_{s,i} = \frac{0.012/0.5}{0.022+0.029} \cdot 1309 = 616$ °C.
(d) Equation 10.40 yields $\tau_f = \frac{0.003 \cdot 560 \cdot 7850}{\frac{1}{0.022+0.029} + \frac{1}{0.049}} = 330$ s and temperature rise after 300 s can be obtained from Eq. 10.39 as $\theta_{core} = 1309 \cdot \frac{0.049}{0.022+0.049+0.029} \cdot \left(1 - e^{-\frac{300}{326}}\right) = 1309 \cdot 0.490 \cdot 0.601 = 385$ °C and from Eq. 10.41 $\theta_f = \frac{0.029 \cdot 1309 + 0.022 \cdot 385}{0.022+0.029} = 910$ °C.
(e) See the plot of Fig. 10.9

Comment: Notice in Fig. 10.9 that according to the theory temperatures, the fire temperature θ_f and the inner exposed surface temperature $\theta_{s,i}$ starts at temperature levels between the initial and the ultimate temperatures depending on the thermal resistances.

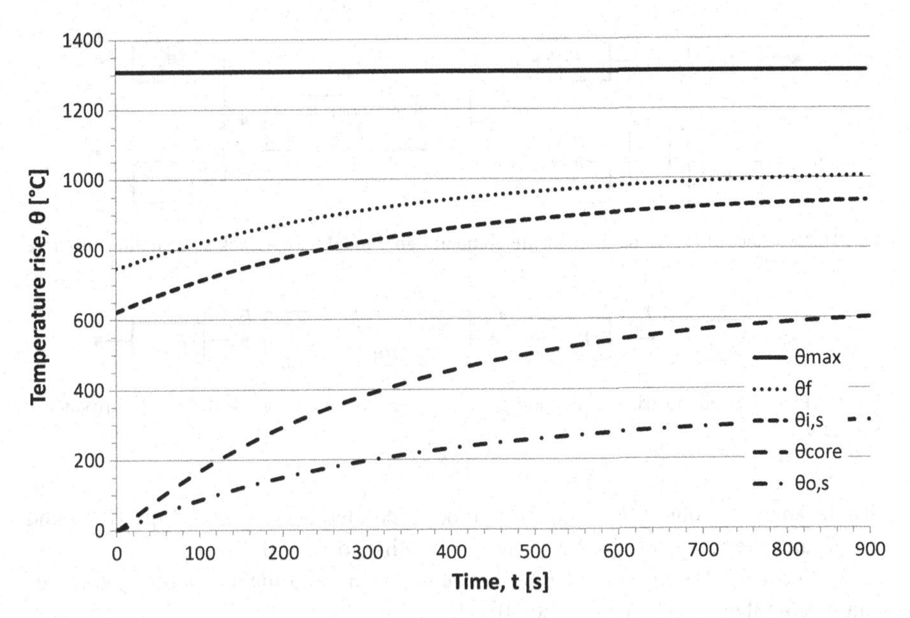

Fig. 10.9 Calculated temperature rises of Example 10.3. Notice that the fire and inner surface temperatures rise instantaneously according to the theory

10.2.3 Temperature-Dependent Material and Heat Transfer Properties: Numerical Solutions

In most cases fire compartments have openings through which heat can radiate out to the environment at ambient temperature. Exceptions are furnaces and tunnels, and therefore tunnel fires may become very hot. The heat losses by radiation out through opening $\dot{q}_r$ according to Eq. 10.8 is small at low temperatures but increases rapidly at elevated temperatures and must therefore be considered particularly when analysing hot fires. As it is highly non-linear as it depends on the fire temperature to the fourth degree. This is also the case for the radiation to the fire compartment surfaces, and in addition the surrounding structure may consist of several layers of materials with properties varying with temperature. Then numerical solutions are required.

In general, Eq. 10.9 is valid as a boundary condition for the one-dimensional model. For a fire compartment with relatively thick but not infinite boundaries, a thermal model as indicated by electric circuit analogy shown in Fig. 10.10 may then be applied.

This model analysed as it is or it may be reduced by the rules of combining resistances and defining combined temperatures to the circuit analogy of Fig. 10.11.

Then the boundary temperature T_{max} is a weighted average value of the ambient temperature T_∞ and T_{ult}. θ_{max} can be calculated by solving Eq. 10.12 for $\dot{q}''_w = 0$.

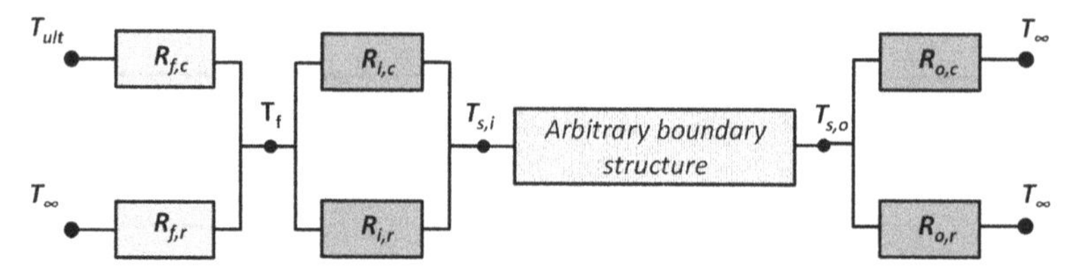

Fig. 10.10 Electric circuit analogy of fire compartment model with a thick surrounding structure

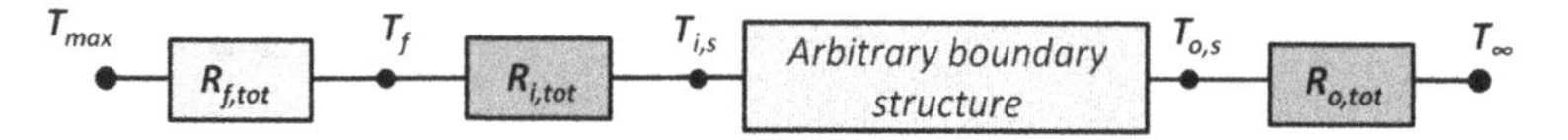

Fig. 10.11 Reduced electric circuit analogy of fire compartment model with a thick surrounding structure

The thermal resistances $R_{f,c}$ and $R_{f,r}$ can be calculated according to Eqs. 10.16 and 10.17, and the resultant resistance $R_{f,tot}$ according to Eq. 10.21.

The heat transfer resistance $R_{i,tot}$ between the fire and the surrounding surfaces can be calculated according to Eq. 10.22.

Now with the boundary condition according to Eq. 10.20 with two heat transfer resistances in series may the temperature in the surrounding structure be calculated including the surface temperature T_s with a general temperature calculation code.

When the surface temperature is calculated, the fire temperature can be obtained as

$$T_f = \frac{R_{f,tot}T_{s,i} + R_{i,tot}T_{max}}{R_{f,tot} + R_{i,tot}} \tag{10.46}$$

Notice that heat transfer resistance due to radiation depends on the temperatures and therefore Eq. 10.46 is implicit and $R_{f,tot}$ and $R_{i,tot}$ must be updated at each time step. Below four cases are shown of calculated and measured fire temperatures in a reduced scale room with dimensions according to Fig. 10.12a. A diffusion propane burner (300 mm by 300 mm) was placed inside the fire compartment releasing a constant power of 1000 kW. It generated immediate flash-over with flames emerging out the door opening, see Fig. 10.12b.

The thermal model as indicated in Fig. 10.10 was analysed with TASEF for the surrounding structures of lightweight concrete and steel sheets. The steel sheets were either insulated on the outside, on the inside or non-insulated. Figure 10.13 shows the measured and calculated temperatures in a compartment of lightweight concrete.

Figure 10.14 shows measured and calculated fire temperatures from the same tests series with a compartment of 3 mm steel sheets insulated on the outside, inside or not at all, respectively. In the calculations the changes of thermal properties of the insulation and the steel were considered.

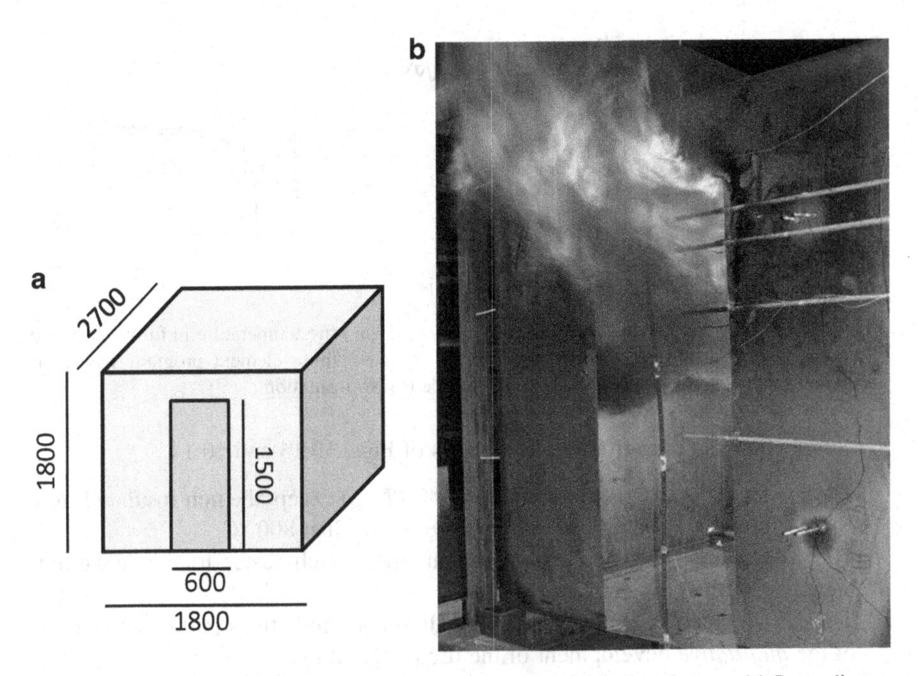

Fig. 10.12 Reduced scale fire compartment experiment with propane burner. (**a**) Inner dimensions (in mm) (**b**) Flames shooting out the door-way

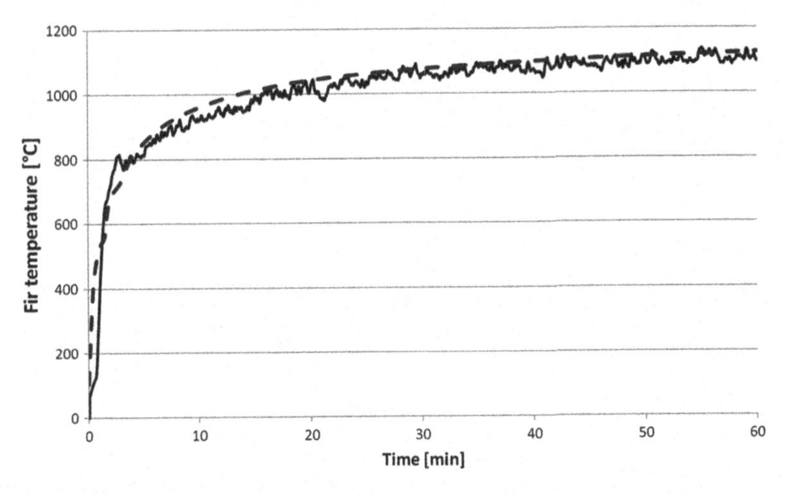

Fig. 10.13 Measured (*full line*) and calculated fire (*dashed line*) temperatures in fully developed compartment fire in a concrete compartment using the finite element program Tasef

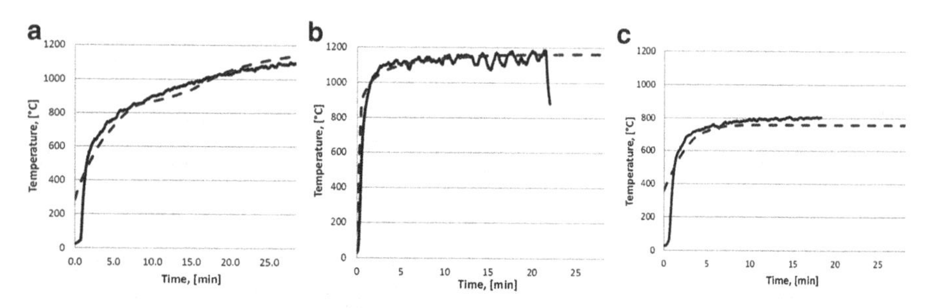

Fig. 10.14 Measured (*full line*) and calculated (*dashed line*) fire temperature in fully developed compartment fire in a steel sheet compartments using the finite element program Tasef. (**a**) Insulation on the outside (**b**) Insulation on the inside (**c**) No insulation

Notice the following from the four cases of Figs. 10.13 and 10.14

- The fire temperature goes to about 1150 °C (T_{max}) except the non-insulated steel compartment where the final temperature is less than 800 °C.
- The inside insulated steel compartment goes much faster to the maximum temperature than the outside insulated.
- The calculation model yields exceptionally good predictions particular in terms of the *qualitative* development of the fire temperature.

Chapter 11
Pre-flashover Compartment Fires: Two-Zone Models

Two-zone models are applied to pre-flashover fires, i.e. compartment fires which have not reached ventilation controlled combustion conditions as defined in Chap. 10. Several more or less advanced computer codes have been developed to calculate temperature under such assumptions. The most fundamental principles of the theory are outlined below.

In most cases the heat release rate as a function of time is input to pre-flashover calculation models. Examples are given in Table 11.1 of the order of magnitude of the heat release rates of various fires.

In the post-flashover model described above the heat release rate was assumed determined by the opening alone, see Eq. 10.3. In pre-flashover models as the one described below the heat release rate $\dot{q}_f$ is an input variable. All combustion is assumed to occur inside the fire compartment boundaries and it is limited by the rate at which gaseous fuel (*pyrolysis gases*) is being released from burning objects. As shown in Fig. 11.1, an upper layer is then supposed to develop where the fire temperature T_f is assumed to be uniform. Below, the lower layer gas temperature remains at the ambient temperature T_∞. Hot combustion gases enter the upper layer by the way of entrainment into the fire plume of flames and combustion gases developed by the burning items. The flow rate $\dot{m}_p$ at which mass is entering the upper layer must balance the mass flows going in $\dot{m}_i$ and out $\dot{m}_o$ of the compartment. Thus

$$\dot{m}_p = \dot{m}_i = \dot{m}_o \tag{11.1}$$

The plume mass flow rate may be calculated as a function of the heat release rate $\dot{q}_f$ and the height of between the fuel surface and the height of the upper layer interface H_D. In the pre-flashover stage of a fire, it is the plume entrainment rate rather than the size of openings that governs the mass flow rate. This is in contrast to post-flashover fires.

U. Wickström, *Temperature Calculation in Fire Safety Engineering*,
DOI 10.1007/978-3-319-30172-3_11

Table 11.1 Examples of the order of magnitude of heat release rates (HRR) of various fires

Item	Typical heat release rate [W]
Wood-burning stove	10
Single burning furniture item	100
Flashover of small room	1000
Full flashover large room/apartment	10,000
Fire in a loaded truck	100,000

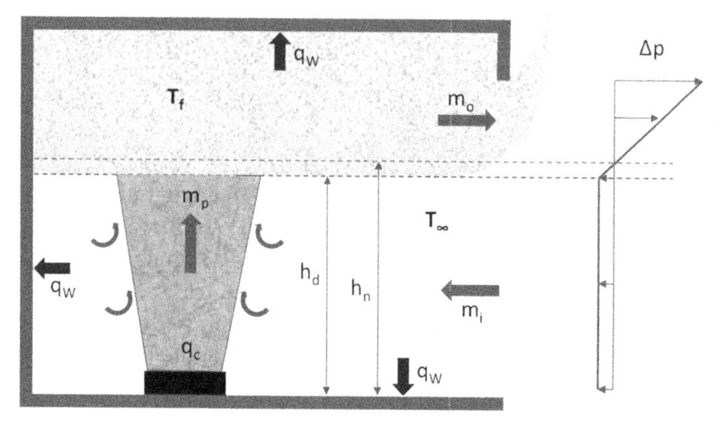

Fig. 11.1 Two-zone model of a pre-flashover room fire with a uniform temperature T_f in the upper layer and ambient temperature T_∞ in the lower layer

As for the one-zone model (post-flashover fires) the heat balance equation of a fire compartment may be written as

$$\dot{q}_c = \dot{q}_l + \dot{q}_w + \dot{q}_r \tag{11.2}$$

where the convection term $\dot{q}_l$ is proportional to the mass flow rate $\dot{m}_p$ and the temperature rise $(T_f - T_\infty)$. In a similar way, the heat loss to the surrounding boundaries $\dot{q}_w$ depends on their thermal properties and the fire temperature. The radiation loss term $\dot{q}_r$ depends on T_f^4 but is less significant for pre-flashover cases as the fire temperature level then in general is lower.

There are two equations, the mass balance and the heat balance, and two unknowns, the temperature T_f and the distance h_D, which now can be solved by a forward time incremental scheme. The input combustion rate $\dot{q}_c$ may vary with time but it is ultimately limited by the availability of oxygen. If too much fuel is released, the fire becomes ventilation controlled and a one-zone model can be assumed, see Chap. 10. Therefore, for two-zone models, cf. Eq. 10.4

$$\dot{q}_c \leq \alpha_1 \alpha_2 A_o \sqrt{h_o} \tag{11.3}$$

When the heat release rate is constant, the temperature rise depends on the thermal properties of the surrounding structure and the ventilation (openings) in a similar way as for the one-zone model.

11.1 Heat and Mass Balance Theory

The plume flow $\dot{m}_p$ may be calculated according to Zukoskis' plume equation [39] as

$$\dot{m}_p = \alpha_3 \, \dot{q}_c^{1/3} \, z^{5/3} \tag{11.4}$$

where z is the effective height of the plume above the burning area. Then with an analogue derivation as for the one-zone case the heat flux to the surrounding structures may be written as

$$\dot{q}''_w = \frac{\alpha_3 \, \dot{q}_c^{1/3} \cdot z^{5/3} \cdot cp}{A_t} \left(\frac{\dot{q}_c^{2/3}}{\alpha_3 \, c_p \, z^{5/3}} - \theta_f \right) + \frac{\varepsilon_f \cdot A_o}{A_t} \sigma \left(T_\infty^4 - T_f^4 \right) \tag{11.5}$$

The emissivity ε_f is here a reduction coefficient considering that the entire opening is not radiating corresponding to the hot zone fire temperature. According to Karlsson and Quintiere [40] $\alpha_3 = 0.0071 \left[\frac{\mathrm{kg}}{\mathrm{W\,s\,m}^{5/3}} \right]$. Now with heat release $\dot{q}_c$ and the effective height of the plume z assumed constant, the fire temperature may be calculated in a similar way as for post-flashover one-zone models.

Thus a resultant temperature rise θ_{max} can be defined as (see Fig. 11.2)

$$\theta_{max} = \frac{R^*_{f,r} \theta^*_{ult}}{R^*_{f,c} + R^*_{f,r}} = \frac{\theta^*_{ult}}{1 + \frac{R^*_{f,c}}{R^*_{f,r}}} \tag{11.6}$$

where θ^*_{ult} is the ultimate gas temperature rise determined as the heat release rate $\dot{q}_c$ over the mass flow rate $\dot{m}_p$ (Eq. 11.4) and the specific heat of air c_p assuming no losses neither through radiation out the openings nor from losses to boundary surfaces. Observe that alternatively θ_{max} can be obtained by solving Eq. 11.5 for θ_f when $\dot{q}''_w$=0.

By comparison with Eq. 11.5, the ultimate temperature can be identified as

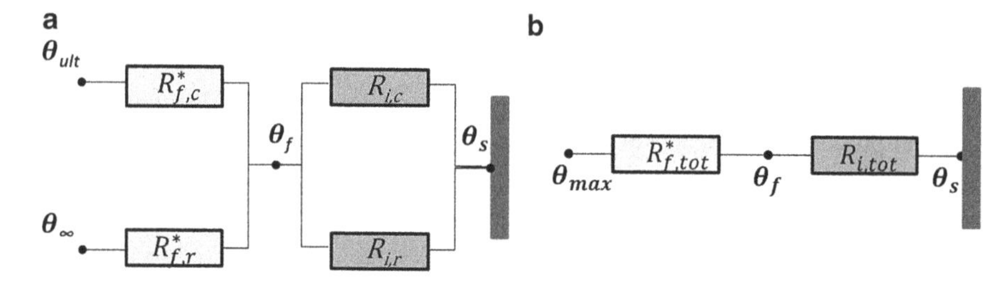

Fig. 11.2 Electric circuit analogy model of a pre-flashover compartment fire boundary. (**a**) Two boundary temperatures (**b**) reduced to one boundary temperature

$$\theta^*_{ult} = \frac{\dot{q}_c^{2/3}}{\alpha_3 \cdot c_p \cdot z^{5/3}} \tag{11.7}$$

and the fire thermal resistances can be identified from as

$$R^*_{f,c} = \frac{A_t}{\alpha_3 \cdot \dot{q}_c^{1/3} \cdot z^{5/3} \cdot c_p} \tag{11.8}$$

and

$$R^*_{f,r} = \frac{A_t}{\varepsilon_f \cdot A_o \sigma \left(T^2_\infty + T^2_f\right) \cdot \left(T_\infty + T_f\right)} \tag{11.9}$$

Then the heat flux to the surface can be calculated as (see Fig. 11.2b)

$$\dot{q}''_w = \frac{\left(\theta_{max} - \theta_f\right)}{R^*_{f,tot}} = \frac{\left(\theta_{max} - \theta_s\right)}{R^*_{f,r} + R_{i,tot}} \tag{11.10}$$

where the resultant resistance

$$R^*_{f,tot} = \frac{1}{\dfrac{1}{R^*_{f,c}} + \dfrac{1}{R^*_{f,r}}} \tag{11.11}$$

Observe that $R^*_{f,r}$ must be calculated based on the fire temperature T_f and on the resultant temperature $T_{max} = \theta_{max} + T_\infty$. In a similar way may resultant heat transfer resistance between the fire gases and compartment boundary surface $R_{i,tot}$ be calculated according to Eq. 10.22 based on the temperatures T_f and the surface temperature T_s.

Then the boundary condition becomes

$$\dot{q}''_w = \frac{1}{R^*_{f,tot} + R_{i,tot}} \left(\theta^*_{max} - \theta_s \right). \tag{11.12}$$

This boundary condition can be used together with thermal model of the surrounding structure in similar way as for post-flashover compartments. Then the surface temperature may be calculated, and thereafter the fire temperature can be obtained by the law of proportion as

$$T_f = \frac{T_s\, R_{f,tot} + T_{max}\, R_{i,tot}}{R_{f,tot} + R_{i,tot}} = \frac{\theta_s\, R_{f,tot} + \theta_{max}\, R_{i,tot}}{R_{f,tot} + R_{i,tot}} \tag{11.13}$$

In general analyses of the boundary structure require numerical methods. However, as for post-flashover one-zone analyses analytical solutions are sometimes possible, given the heat resistances are given constant values representing a relevant temperature level.

11.2 Solution of the Upper Layer Fire Temperature

In combination with the boundary condition as defined by Eq. 11.12 may the temperature of a surrounding structure be calculated in a similar way as for one-zone models. Thus based on the calculated fire-exposed surface temperature may then the upper layer temperature be calculated. In the next two sections will the cases of assumed semi-infinite and thin structures, respectively, be analysed. As for one-zone models analytical solutions may under certain conditions be derived for quick rough estimates.

11.2.1 Semi-infinitely Thick Compartment Boundaries

The surface temperature of semi-infinitely thick compartment boundaries with boundary conditions according to Eq. 11.12 may of course be solved numerically with, e.g. finite element methods. However, with the assumptions of

- Constant heat release rate
- Constant material properties, $(k \cdot \rho \cdot c)$
- Constant thermal heat transfer coefficients/resistances

may a closed form solution be derived as for one-zone models (see also Sect. 3.2.3), i.e.

$$\theta_s = \theta_{ult}^* \left[1 - e^{\frac{t}{\tau_f^*}} erfc\left(\sqrt{\frac{t}{\tau_f^*}}\right)\right] \quad (11.14)$$

the time constant τ_f^* can be calculated as

$$\tau_f^* = \frac{k \cdot \rho \cdot c}{\left(\frac{1}{R_{f,c}^* + R_{i,tot}^*}\right)^2} = k \cdot \rho \cdot c \cdot \left(R_{f,c}^* + R_{i,tot}\right)^2 \quad (11.15)$$

and then Eq. 10.23 yields the fire temperature.

The theory as outlined here gives interesting qualitative results but needs to be further validated by comparing with well-controlled experiments. Some such comparisons have been done with very good results for compartments where the heat capacity of the boundary can be lumped into a steel sheet as is shown in the next section. Below is an example with purpose of showing how fire temperatures can be calculated.

Example 11.1 A propane gas burner at a height of 0.5 m in the room/corner test room was set at a constant power 450 kW. The room has a total surrounding area $A_t = 44$ m^2 and door opening $A_O = 2$ m^2. Assume effective height of the burner plume $z = 1$ m. Assume all the surrounding structural elements being infinitely thick light-weight concrete with a thermal inertia $k \cdot \rho \cdot c = 0.2 \cdot 500 \cdot 800 = 80 \cdot 10^3\, \mathrm{W^2\, s/(m^4\, K^2)}$. Initial and ambient temperatures are equal to 20 °C.

(a) Calculate the maximum temperature *not* considering the radiation out the door-way.
(b) Calculate the maximum temperature considering the radiation out the door-way.
(c) Derive the surface and the fire temperatures as functions of time *not* considering the radiation out the door-way and calculate the surface and fire temperatures after 15 min. Assume a constant heat transfer coefficient $h_i = 25$ W/(m K).

Solution

(a) After a long time the wall losses vanish and the maximum temperature rise can be derived from Eq. 11.7: $\theta_f = \theta_{ult} = \frac{\dot{q}_c^{2/3}}{\alpha_3 \cdot c_p \cdot z^{5/3}} = \frac{450000^{2/3}}{0.0071 \cdot 1150 \cdot 1^{5/3}}$ $= 722\,\mathrm{K}$. Hence the maximum temperature $T_f = 722 + 20 = 724\,°\mathrm{C}$.

(b) Then $\dot{q}_w = \frac{\alpha_3 \cdot \dot{q}_c^{1/3} \cdot z^{5/3} \cdot c_p}{A_{tot}} \left(\frac{\dot{q}_c^{2/3}}{\alpha_3 \cdot c_p \cdot z^{5/3}} - \theta_f\right) + \frac{A_o}{A_{tot}} \cdot \sigma \cdot \left(T_\infty^4 - T_f^4\right) =$ $\frac{0.0071 \cdot 450000^{1/3} \cdot 1150}{44} \cdot (722 - \theta_f) + \frac{2}{44} \cdot 5.67 \cdot 10^{-8} \cdot \left[(273+20)^4 - (\theta_f + 273 + 20)^4\right] = 0$. This 4th degree equation yields a temperature rise $\theta_f = 615\,\mathrm{K}$ and $T_f = 625\,°\mathrm{C}$.

(c) The fire temperature as a function may be obtained from Eq. 11.14. From Eq. 11.8 $R^*_{f,c} = \frac{A_{tot}}{\alpha_3 \cdot \dot{q}_c^{1/3} \cdot z^{5/3} \cdot c_p} = \frac{44}{0.0071 \cdot 450000^{1/3} \cdot 1^{5/3} \cdot 1150}$ $= 0.0704 \quad (\text{m K})/\text{W}$ and $\theta^*_{ult} = 722\,\text{K}$. Equation 11.15 yields $\tau^*_f = \frac{80 \cdot 10^3}{\left(\frac{1}{0.0704 + \frac{1}{25}}\right)^2} = 975\,\text{s}$. Thus according to Eq. 11.14 $\theta_s = 722 \cdot \left[1 - e^{\left(\frac{t}{975}\right)} \cdot erfc\left(\sqrt{\frac{t}{975}}\right)\right]$. After 900 s $t/\tau^*_f = 900/975 = 0.92$. The value of the function between the brackets from Fig. 3.11 or Table 3.3 is 0.56. Then $\theta_s = 722 \cdot 0.56 = 404\,\text{K}$. The fire temperature can then be obtained from Eq. 11.13 as a mean weighted value of the surface and the ultimate temperature as $\theta_f = \frac{404 \cdot 0.0704 + 722 \cdot \frac{1}{25}}{0.0704 + \frac{1}{25}} = 519\,\text{K}$ and the fire temperature $T_f = 529\,°\text{C}$.

11.2.2 Insulated and Uninsulated Boundaries with a Metal Core

With the same assumptions as specified in Sect. 10.2.2 a similar expression as for one-zone models can be obtained for the core temperature.

Referring to Fig. 10.8 and Table 10.2 for the definitions of the parameters, the core temperature can be numerically solved by the forward difference recursion formula (c.f. Eq. 10.38)

$$\theta^{i+1}_{core} = \theta^i_{core} + \frac{\Delta t}{C_{core}} \left[\left(R^*_{f,tot} + R_{i,tot} + R_{i,ins} \right) (\theta_{max} - \theta_{core}) - (R_{o,ins} + R_{o,tot}) \cdot (\theta_{core} - \theta_\infty) \right] \quad (11.16)$$

where θ_{max} is defined by Eq. 11.6, $R^*_{f,tot}$ Eq. 11.8 and the other parameters as in Sect. 10.2.2.

As for post-flashover fires (one-zone models) an analytical solution may be derived if the heat transfer parameters and the material properties are assumed constant, not changing with temperature. Thus (cf. Eq. 10.39)

$$\theta_{core} = \theta^*_{ult} \left[\frac{R_{o,tot} + R_{o,ins}}{R^*_{f,c} + R_{i,tot} + R_{i,ins} + R_{o,tot} + R_{o,ins}} \right] \left(1 - e^{-t/\tau^*_f} \right) \quad (11.17)$$

and the fire temperature rise as (cf. Eq. 10.41)

$$\theta_f = \frac{\theta_{ult}^*}{1 + \frac{R_{i,tot} + R_{i,ins}}{R_{f,c}^*}} \left[\frac{R_{o,tot} + R_{o,ins}}{R_{f,c}^* + R_{i,tot} + R_{i,ins} + R_{o,tot} + R_{o,ins}} \left(1 - e^{-t/\tau_f^*}\right) + \frac{R_{i,tot} + R_{i,ins}}{R_{f,c}^*} \right] \tag{11.18}$$

where the time constant is calculated as

$$\tau_f^* = \frac{C_{core}}{\frac{1}{R_{f,c}^* + R_{i,tot} + R_{i,ins}} + \frac{1}{R_{o,tot} + R_{o,ins}}} \tag{11.19}$$

Equation 11.18 yields a crude estimate of the fire temperature development as several assumptions are made to linearize the problem. More accurate solutions can be made by the step-by-step numerical procedure according to Eq. 11.16 whereby the material and, in particular, heat transfer conditions can be updated at each time step. Such a calculation procedure was implemented in an MS-Excel sheet by Evegren and Wickström [41]. It was used to predict and compare with measured temperatures in an uninsulated and an insulated steel container with a burning pool of heptane, see Fig. 11.3.

Fig. 11.3 Dimensions of the test enclosure and photo of the insulated test enclosure and the pool fire experiment. From [41]

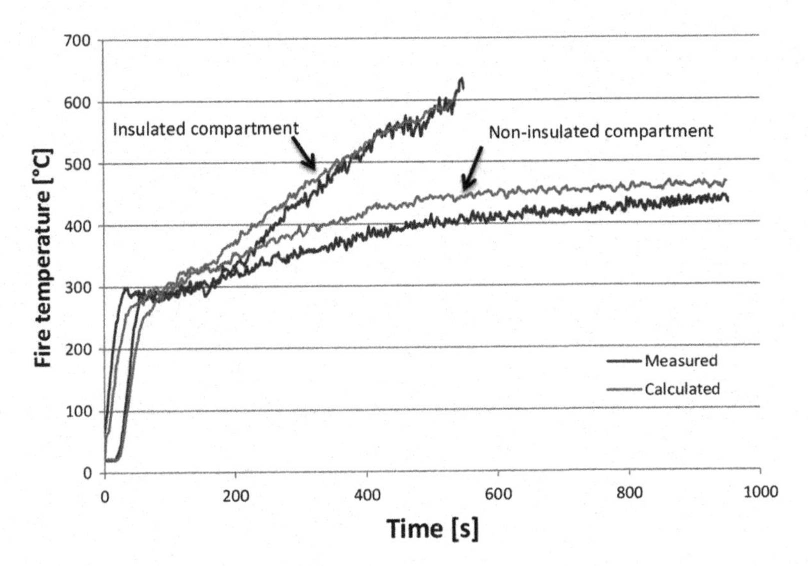

Fig. 11.4 Measured and calculated upper layer fire temperature. From [41]

The results are shown in Fig. 11.4. As can be seen predictions were very accurate for both the cases.

Chapter 12
Fire Exposure of Structures According to Standards

When exposed to fire structures deform and lose load-bearing capacity which must be considered in design processes. It is then exposures to the more severe fires which are of interest such as post-flashover compartment fires and large flames for longer times. Pre-flashover fires do in general not create thermal conditions that can jeopardize the function of structural elements in a building. For design purposes it is therefore in general exposures relevant for post-flashover compartment fires that are specified in various standards and guidelines in the form of time–temperature curves. These curves are then used for controlling fire resistance test furnaces, see Fig. 12.1.

They can also be used as fire temperatures when predicting temperature of structures exposed to standard fire conditions. When predicting test according to the international standard ISO 834 and the European standard EN 1363-1 the gas temperature and the radiation temperature may be assumed equal as these standards prescribe plate thermometers for controlling of furnace temperature. However, when predicting tests according to the American standard ASTM E-119, deviations due to the thick thermocouples specified for controlling the furnace temperature should be considered, see Sect. 9.1.3.

A deterministic design and analysis process of structures exposed to fire entails three major steps:

1. Determine the fire exposure to which the surface of the structure is subjected.
2. Determine the thermal response of the structure to the exposing fire.
3. Determine the structural response and the load-bearing capacity at elevated temperatures.

This chapter is focusing on the second step. The first section deals with design fires then followed by sections on the structural materials concrete, steel and wood.

U. Wickström, *Temperature Calculation in Fire Safety Engineering*,
DOI 10.1007/978-3-319-30172-3_12

Fig. 12.1 A glazed partition being tested in a vertical fire resistance furnace. Notice the Plate thermometers for monitoring the furnace temperature

12.1 Standard Time Temperature Fire Curves

The so-called standard fire curve as defined in the European standard EN 1363-1 and the international standard ISO 834 is outside the USA and Canada the by far most commonly used time–temperature relation used for testing and classification of separating and load-bearing building structures. The time–temperature relation of this *EN/ISO standard fire temperature* curve is then specified as

$$T_f = 20 + 345 \cdot \log(8t + 1) \tag{12.1}$$

where T_f is temperature in °C and t is time in minutes. A selection of time temperature coordinates is given in Table 12.1.

A so-called *external fire curve* is given in Eurocode 1 (EN 1991-1-2) as

$$T_f = 20 + 660 \cdot \left(1 - 0.687\, e^{-0.32 \cdot t} - 0.313\, e^{-3.8 \cdot t}\right) \tag{12.2}$$

This time–temperature is intended to be used for external structures outside of external walls.

When more severe fires are anticipated, as for offshore oil installations or tunnels, the so-called *Hydrocarbon Curve* is often applied:

Table 12.1 Time–temperature coordinates of the standard ISO 834 and EN 1363-1 fire curves

Time [min]	0	15	30	45	60	90	120	180	240
Temperature [°C]	20	739	842	2	945	1006	1049	1110	1153

Table 12.2 Time–temperature coordinates of the RWS fire curve

Time [min]	0	3	5	10	30	60	90	120	180
Temperature [°C]	20	890	1140	1200	1300	1350	1300	1200	1200

$$T_f = 20 + 1080 \cdot \left(1 - 0.325\, e^{-0.167 \cdot t} - 0.675\, e^{-2.5 \cdot t}\right) \tag{12.3}$$

For the design of tunnels, the ministry of transport in the Netherlands has developed the so-called *RWS fire curve* which is used in many countries. It is defined by the time–temperature coordinates given in Table 12.2.

The above-mentioned fire design curves are plotted in Fig. 12.2 together with time–temperature curve according to ASTM E-119.

In the USA and Canada fire tests and classification are generally specified according to the standard ASTM E-119. The *ASTM E-119 fire curve* is specified as time temperature coordinates as given in Table 12.3 or as approximated by the equation

$$T_f = T_0 + 750\left(1 - e^{-0.49\sqrt{t}}\right) + 22.0\sqrt{t} \tag{12.4}$$

where fire or furnace temperature T_f and the initial temperature T_0 are in °C and time t in minutes. The curve is slightly different from the corresponding ISO and EN curves. However, the severity of a fire test depends not only on temperature level but also on how the temperature is measured. In ASTM E-119 the thermocouples specified for monitoring the furnace temperature are very thick and have therefore a very slow response. That means the real temperature level is much higher than measured by thermocouples. During the first 10 min of a fire resistance test, the difference between measured temperature and the actual temperature level may amount to several hundred degrees as indicated in Fig. 9.1. Thus, when predicting temperature in structures to be tested according to ASTM E-119, the most relevant fire temperature curve to apply is the upper curve of Fig. 9.1. In addition to the problem of the time constant, it is unclear how the ASTM thermocouples react to different gas and radiation temperatures which makes any temperature predictions uncertain.

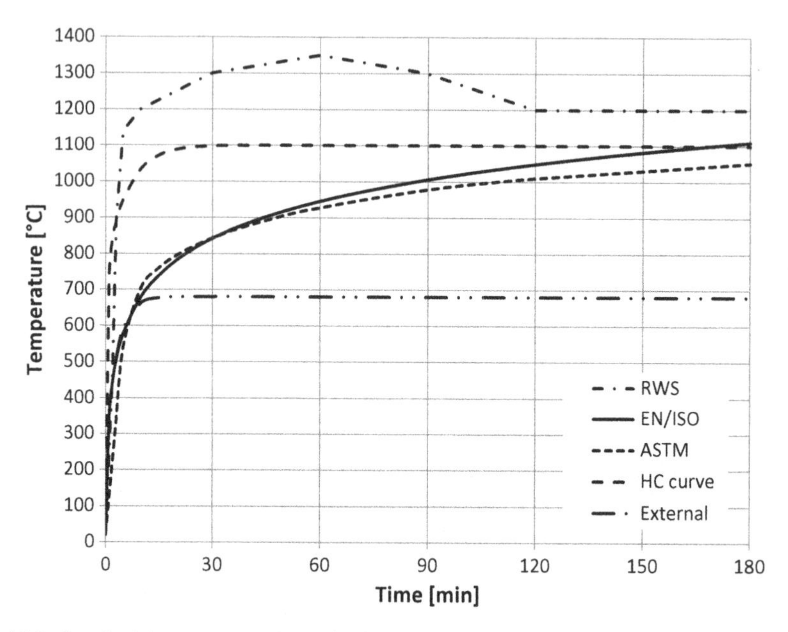

Fig. 12.2 Standard time–temperature relations according to ISO 834 or EN 1363-1 (Eq. 12.1), the Hydrocarbon curve (Eq. 12.2), the External fire curve according to Eurocode 1 (Eq. 12.3) and ASTM E-119 (defined in Table 12.3 and approximated by Eq. 12.4)

Table 12.3 Time–temperature coordinates of the ASTM E-119 fire curve

Time [min]	0	5	10	30	60	120	240	480	>480
Temperature [°C]	20	538	704	843	927	1010	1093	1260	1260

12.2 Parametric Fire Curves According to Eurocode

Parametric fire curves are defined in Eurocode 1, EN1991-1-2, Appendix A. They are based on work in Sweden [36] in the 1960s and 1970s which was later modified and simplified by Wickström, see e.g. [37].

The parametric fire curves are defined in the heating phase by the expression

$$T_f = 20 + 1325 \cdot \left(1 - 0.324 \cdot e^{-0.2 \cdot t^*} - 0.204 \cdot e^{-1.7 \cdot t^*} - 0.472 \cdot e^{-19 \cdot t^*}\right) \quad (12.5)$$

where t^* is a modified time defined as

$$t^* = \Gamma \cdot t \quad (12.6)$$

where the parameter Γ (*the gamma factor*) determines the rate at which the fire temperature goes to the ultimate temperature, (1325 + 20) °C. For Γ equal unity Eq. 12.5 yields a time–temperature relation which approximately follows the standard EN/ISO curve for about 6 h. The standard fire curve prescribes thereafter higher temperatures while the parametric goes asymptotically a maximum fire temperature rise of 1345 °C.

The factor Γ depends on the opening factor (cf. Eq. 10.10) and the thermal inertia of surrounding structures. It is defined as

$$\Gamma = \left[\frac{\frac{A_o\sqrt{h_o}/A_t}{\sqrt{k\cdot\rho\cdot c}}}{\frac{0.04}{1160}} \right]^2 = 841 \cdot 10^6 \cdot \frac{O^2}{k \cdot \rho \cdot c}. \quad (12.7)$$

(The parameter values in this equation must be given in SI units.) Thus a fire compartment with an opening factor of $O = 0.04\ \mathrm{m}^{1/2}$ and enclosure boundaries with a characteristic value of the square root of the thermal inertia $\sqrt{k \cdot \rho \cdot c} = 1160$ $\mathrm{W\,s}^{1/2}/(\mathrm{K\,m}^2)$ yields $\Gamma = 1$ which implies a fire development close to the EN/ISO standard fire curve. Lower values, $\Gamma < 1$, yield fires with slower temperature developments while $\Gamma > 1$ yields fires with faster developments.

Figure 12.3 shows examples of the heating phase of parametric fire curves with Γ-values smaller and larger than unity. The ISO/EN standard curve is plotted for comparison. Notice that the parametric fire curve with $\Gamma = 1$ differs only a few degrees from the ISO/EN standard curve.

The so-called hydrocarbon fire curve is a special case of a parametric curve. It was originally defined as a parametric fire curve with $\Gamma = 50$ and with an ultimate temperature of 1100 °C, see Eq. 12.3.

Fires are assumed to continue until all the fuel (fire load) is consumed, and the burning rate is assumed to be proportional to the amount of air being available in the fire compartment. Thus the *fire duration* t_d is proportional to the fire load density q_f'' (energy content per unit area) and the inverse of the opening factor. According to Eurocode 1 it may be written as

$$t_d = \chi \frac{q_f'' \cdot A_t}{A_o\sqrt{h_o}} \quad (12.8)$$

In modified time the fire duration is calculated as

$$t_d^* = \Gamma \cdot t_d \quad (12.9)$$

In Eurocode 1 the proportionality constant has been given the value $\chi = 0.2{*}10^{-3}$ $\left[\mathrm{h \cdot m^{3/2}/MJ}\right]$ (units as in Eurocode 1 [35]).

The fire load density q_f'' is obtained by summarizing the weight of the various fuel components available for combustion with their net calorific value. Table 12.4 shows a summary of the net calorific values as given in Eurocode 1.

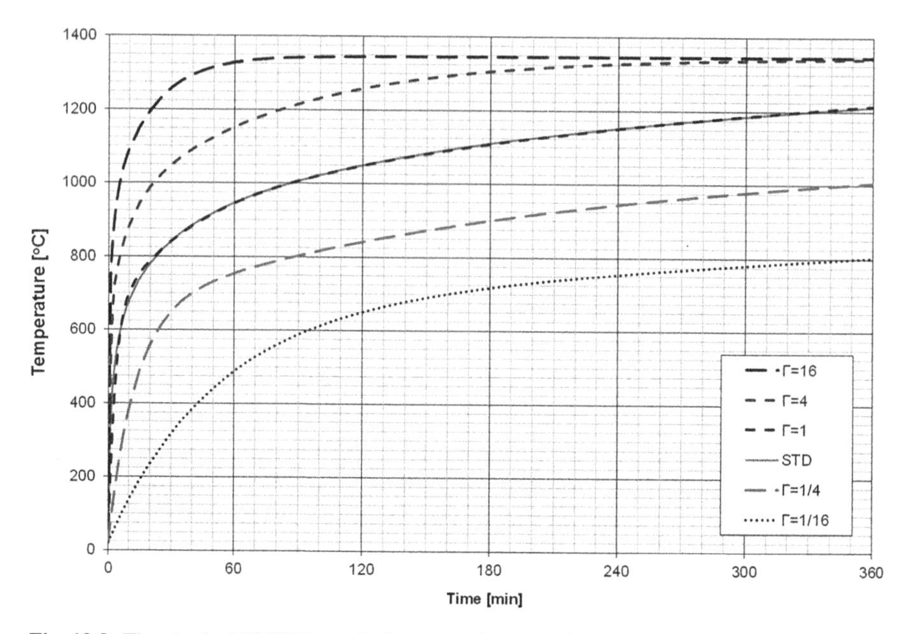

Fig. 12.3 The standard EN/ISO standard curve and parametric fire curves with various Γ-values. For $\Gamma = 1$ the parametric curve coincides approximately with the standard curve for the first 360 min

Table 12.4 Net calorific values of combustible materials for calculation of fire loads

Material	Net calorific values [MJ/kg]
Wood	17.5
Other cellulosic materials	20
Gasoline, petroleum	45
Diesel	45
Polyvinylchloride, PVC (plastic)	20
Other plastics	30–40
Rubber tyre	30

Note: The values given in this table are not applicable for calculating energy content of fuels Summary from Eurocode 1, EN 1991-1-2 (The net calorific value is determined by subtracting the heat of vapourization of the water vapour from the gross calorific value)

The simple heat and mass balance theory applied for calculating the compartment fire temperature in the heating phase is not relevant for the *cooling phase* when the fuel is more or less depleted and the assumption of uniform temperature is no more relevant. In Eurocode 1 simple linear time temperature relations are therefore assumed as shown below:

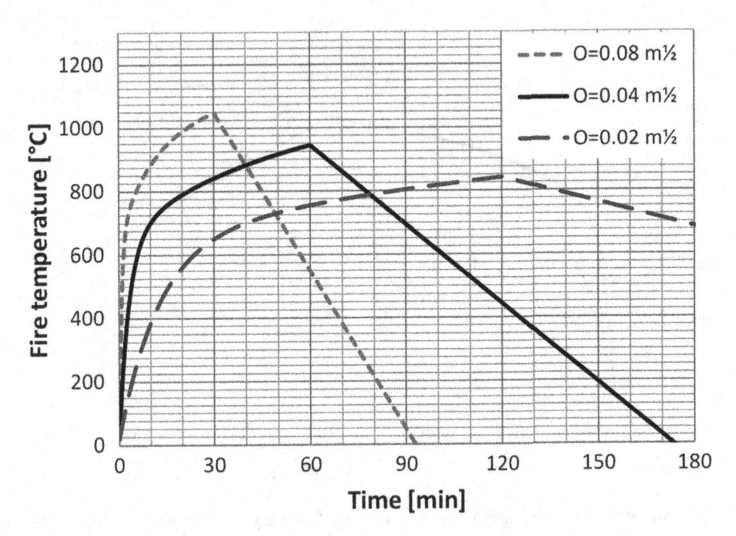

Fig. 12.4 Parametric time–temperature fire curves for varying opening factors O with a fire load $q_f'' = 200\,\mathrm{kJ/m^2}$ and thermal inertia of surrounding boundaries $\sqrt{k\rho c} = 1160\,\mathrm{W\,s^{1/2}/(Km^2)}$. $O = 0.04\ \mathrm{m^{½}}$ yields approximately the ISO standard curve in the heating phase (The curves have Γ-factors 4.0, 1.0 and 0.25, respectively)

$$\begin{aligned} T_f &= T_{f,max} - 625\left(t^* - t_d^*\right) && \text{for} \quad t_d^* \leq 0.5\,\mathrm{h} \\ T_f &= T_{f,max} - 250\left(3 - t_d^*\right)\left(t^* - t_d^*\right) && \text{for} \quad 0.5 < t_d^* \leq 2\,\mathrm{h} \\ T_f &= T_{f,max} - 250\left(t^* - t_d^*\right) && \text{for} \quad t_d^* > 2\,\mathrm{h} \end{aligned} \tag{12.10}$$

In case of fire durations $t_d < 25\,\mathrm{min}$ additional information need to be considered as given in EN 1991-1-2.

Figures 12.4 and 12.5 show examples of parametric fire curves. In both cases the fire load $q_f'' = 300\,\mathrm{kg/m^2}$. Figure 12.4 shows the temperature development for various opening factors and Fig. 12.5 for various thermal inertia $\sqrt{k \cdot \rho \cdot c}$. The Γ-factors 4.0, 1.0 and 0.25 are calculated based on Eq. 12.7.

Notice in Fig. 12.4 how the maximum fire temperature $T_{f,max}$ increases with O while the fire duration t_d decreases. Thus, e.g. concrete structures with slow time responses are in general more sensitive to fires with low opening factor while it is the opposite for bare steel structures.

Figure 12.5 shows that fire temperature development depends on the thermal inertia of the surrounding boundaries. The fire duration t_d depends, however, only on the fire load and the opening factor but is independent of the thermal inertia. Notice that the temperature development is much lower for a thermal inertia of 2000 W $\mathrm{s^{1/2}/(K\ m^2)}$ corresponding to concrete than for approximately the ISO standard curve, $\sqrt{k\rho c} = 1160\,\mathrm{W\,s^{1/2}/(Km^2)}$. After 90 min the difference is. On the other hand, the temperature becomes much higher if the thermal inertia of the surrounding structure is lower, e.g. with a thermal inertia of 500 $\mathrm{Ws^{1/2}/(Km^2)}$ representing silicate or gypsum boards.

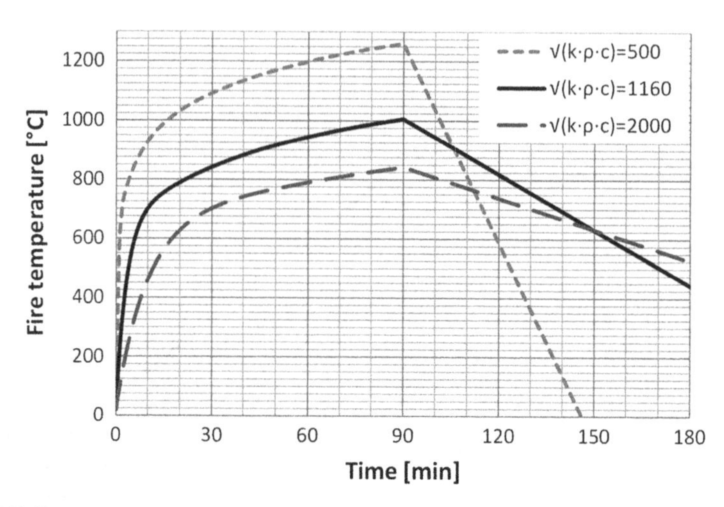

Fig. 12.5 Parametric time–temperature fire curves for varying thermal inertia of surrounding boundaries $\sqrt{k\rho c}$ with a fire load $q_f'' = 200\,\mathrm{kJ/m^2}$ and an opening factor $O = 0.04\ \mathrm{m}^{1/2}$. $\sqrt{k\rho c} = 1160\,\mathrm{W\,s^{1/2}/(Km^2)}$ yields approximately the ISO standard curve in the heating phase (The curves have Γ-factors 4.0, 1.0 and 0.25, respectively)

Example 12.1 Calculate the maximum temperature of a parametric fire in a compartment with $O = A_o\sqrt{h_o}/A_t = 0.08\,\mathrm{m}^{1/2}$, $\sqrt{k \cdot \rho \cdot c} = 1160\,\mathrm{W\,s^{1/2}/(K\,m^2)}$ and $q_f'' = 400\,\mathrm{MJ/m^2}$. In Eq. 12.8 assume $\chi = 0.2{*}10^{-3}\,\mathrm{h\,m^{3/2}/MJ}$.

Solution Equation 12.7 yields $\Gamma = 2^2 = 4$ and Eq. 12.8 $t_d = 0.2{*}10^{-3}{*}400\,/0.08 = 1\,\mathrm{h}$. The maximum temperature can then be obtained for a modified fire duration of $4 \cdot 1$ h = 4 h from Eq. 12.5 or from Fig. 12.3 to be 1150 °C.

12.3 Summary of Heat Transfer Conditions According to Eurocodes

The temperature shall be measured with PT's (see Sect. 9.3) in fire resistance furnace tests according to ISO or EN standards. Therefore the standard furnace temperature T_f can be considered as an AST and the heat transfer $\dot{q}''_{tot}$ to an exposed surface with a temperature T_s may be calculated as (cf. Eq. 4.14)

$$\dot{q}''_{tot} = \varepsilon \cdot \sigma \left[(T_f + 273)^4 - (T_s + 273)^4 \right] + h_c (T_f - T_s) \qquad (12.11)$$

where temperatures here are in °C. This does not apply to predictions of tests according to ASTM E-119 where the fire time–temperature curve needs be modified before being used as a boundary condition in temperature calculations, see Sect. 12.1 above.

When calculating temperatures of structures exposed to severe fires, the influence of the choice of the emissivity ε is in general small and of the convective heat transfer coefficient h_c it is often negligible for most materials, especially light materials, and longer fire durations. For insulation materials (low thermal inertia) it is almost negligible. Then the surface temperature may be assumed equal to the fire temperature (first kind of boundary condition, see Sect. 9.3). On the contrary, for bare/non-protected steel structures particularly the emissivity ε and in some cases also the convection heat transfer coefficient h_c have a significant influence and are decisive for the temperature development.

The emissivity ε is a property of the solid surface only, while the convective heat transfer coefficient h_c depends on the geometry and the surrounding flow conditions. According to Eurocode 1 [35] the specimen surface emissivity $\varepsilon = 0.8$ unless another value can be motivated. The convection heat transfer coefficient $h_c = 25\,\mathrm{W/(m^2\,K)}$ when applying the EN 1363-1 standard curve or the external fire curve, and $h_c = 50\,\mathrm{W/(m^2\,K)}$ and $h_c = 35\,\mathrm{W/(m^2\,K)}$ when applying the hydrocarbon curve or any natural fire curve including parametric fire curves, respectively.

At surfaces on the unexposed side of separating elements Eurocode 1 [35] suggests that the heat transfer shall be calculated as (here the emissivity and view factor of the fire are assumed to be unity)

$$\dot{q}''_{tot} = \varepsilon \cdot \sigma\left[(T_\infty + 273)^4 - \left(T_s + 273^4\right)\right] + h_c(T_\infty - T_s) \qquad (12.12)$$

where T_∞ is the ambient surrounding temperature. Here the unexposed side surface emissivity should be as for the exposed side, i.e. $\varepsilon = 0.8$ unless another value can be motivated. The convection heat transfer coefficient shall then be assumed as $h_c = 4\,\mathrm{W/(m^2\,K)}$. Alternatively, the radiation component of Eq. 12.12 may be included in the convection heat transfer term. Then the convection heat transfer coefficient $h_c = 9\,\mathrm{W/(m^2\,K)}$ and the surface emissivity $\varepsilon = 0$, i.e. a linear boundary condition of the kind 3a according to Table 4.2.

A summary of the heat transfer parameters as specified by Eurocode 1 is given in Table 12.5.

Table 12.5 Summary of heat transfer parameters as specified by Eurocode 1, EN 1991-1-2

Fire curve	Exposed side	Convection heat transfer coefficient h_c [W/(m K)]	Emissivity ε [–]
Standard ISO/EN	Fire-exposed side	25	0.8[a]
	Unexposed side[b]	4	0.8[a]
	Unexposed side[b]	9	0
Hydrocarbon curve	Fire-exposed side	50	0.8[a]
External fire curve	Fire-exposed side	25	0.8[a]

[a]Unless another material property value is motivated
[b]Alternative

Chapter 13
Temperature of Steel Structures

Steel is sensitive to high temperature. The *critical temperature* of a steel member is the temperature at which it cannot safely support its load.

The mechanical properties such as strength and modus of elasticity deteriorate in particular when the steel temperature exceeds 400 °C, see, e.g. Eurocode 3, EN 1993-1-2. Some building codes and structural engineering standard practice defines different critical temperatures which must not be exceeded when exposed to a standard fire exposure for a specified time. Steel structures must therefore usually be protected to reach a particular fire rating. Please note that *insulation* and *protection* of structures are in this book used synonymously. Protections can be obtained by for instance boards, sprayed on concrete, insulation materials or intumescent paint. Intumescent coatings or reactive coatings expand upon heating and provide an insulating char to protect structural steelwork. Steel structures may also be built into concrete or even wooden structures as a means of fire protection. Eurocode 4 (EN 1994-1-2) deals with composite structures of steel and concrete where steel sections are imbedded in concrete.

To obtain a certain rating a steel structure can be tested in fire resistance furnace according to specific standards depending on country or region. Alternatively or as a pretest investigation steel temperatures can be calculated compared with critical values when exposed to design fire conditions for specified durations.

Because of the high conductivity the temperature field in a steel section is in many fire engineering cases assumed uniform. In particular the temperature across the thickness of a steel sheet can in almost all fire resistance cases be assumed constant, while the temperature in the plane of steel sheets may vary considerably. Then the zero- or one-dimensional calculation techniques may be used as presented in Sects. 3.1 and 7.1 and further adapted to *protected* and *unprotected* steel sections in Sects. 13.3 and 13.4, respectively. For more general two- and three-dimensional cases numerical computer codes are needed, see Sect. 7.3.2 and Sect. 13.5 where some examples are shown.

Generally in the following sections the gas and radiation temperatures are assumed equal to the fire temperature, i.e. $T_g = T_r = T_f$, as is assumed in all

U. Wickström, *Temperature Calculation in Fire Safety Engineering*,
DOI 10.1007/978-3-319-30172-3_13

standard time–temperature fire curves. If measured temperatures are used, the fire temperature T_f may be replaced by the adiabatic surface temperature measured with, e.g. PTs.

13.1 Thermal Properties of Steel

Metals in general have high electric conductivity, high thermal conductivity and high density. The heat conductivity of carbon steel is in the order of 30 times higher than the corresponding value for concrete and 100–1000 times higher than that of insulation products. The higher purity of a metal, the better it conducts heat. Thus contents of carbon and alloying metals such as chrome reduce the conductivity, and consequently stainless steel is a relatively poor conductor. The specific heat capacities of metals are in accordance with a general rule of physics inversely proportional to the molecular weight.

Figure 13.1 shows the *conductivity* k_{st} vs. temperature T_{st} of *structural carbon steel* according to Eurocode 3 (EN 1993-1-2). It can also be obtained from Table 13.1. For approximate calculations normally on the safe side a constant value of 46 W/(m K) can be recommended, cf. Table 1.2.

The *specific heat capacity* is usually a more significant parameter than the conductivity for the development of temperature in fire-exposed steel structures. In many cases it is accurate enough and convenient to assume a constant specific heat capacity. Then a value of 460 J/(kg K) is recommended which normally yields calculated temperatures on the safe side (overvalued). However, for more accurate calculations the variations with temperature as shown in Fig. 13.2 or given in Table 13.2 are recommended in Eurocode 3 [3]. The peak of the specific heat capacity at 735 °C is due to phase changes of the steel.

Table 13.3 shows tabulated values of the thermal properties of carbon steel derived from Eurocode 3 including the specific volumetric enthalpy vs. temperature defined as

$$e(T) = \int_0^T c \cdot \rho \, dT \tag{13.1}$$

This temperature–enthalpy relation is input in some computer codes, e.g. Tasef, instead of density and specific heat capacity. The diagram in Fig. 13.3 shows the specific volumetric enthalpy vs. temperature based on the values of Table 13.3.

Thermal conductivity of *stainless steel* is considerably lower than that of carbon steel. The conductivity and the specific heat capacity of stainless steel according to Eurocode 3, EN 1993-1-2 are given in Table 13.4.

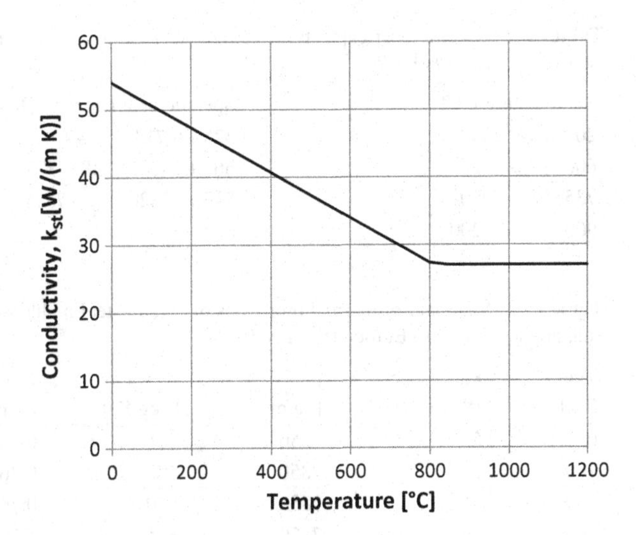

Fig. 13.1 Thermal conductivity of steel vs. temperature according to Eurocode 3, EN 1993-1-2. See also Table 13.1

Table 13.1 Thermal conductivity of carbon steel vs. temperature according to Eurocode 3, EN 1993-1-2

Temperature [°C]	Conductivity [W/m K]
$20 < T_{st} < 800$	$54 - 0.0333\, T_{st}$
$800 < T_{st} < 1200$	27.3

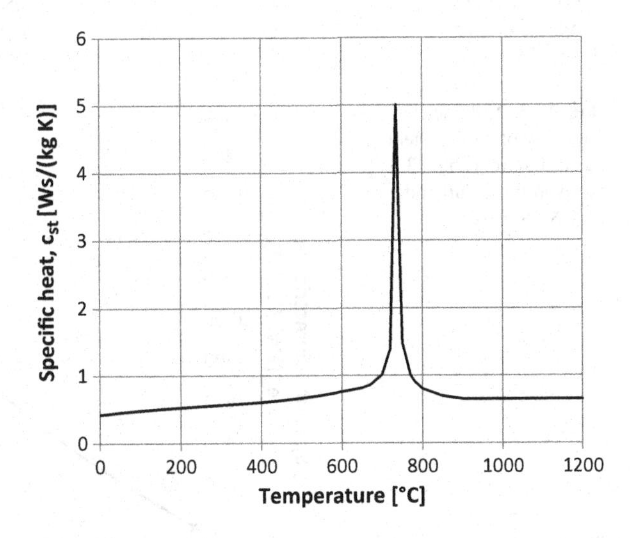

Fig. 13.2 Specific heat capacity of carbon steel vs. temperature according to Eurocode 3, EN 1993-1-2

Table 13.2 Specific heat capacity of carbon steel as functions of the temperature according to Eurocode 3, EN 1993-1-2

Temperature [°C]	Specific heat capacity [J/(kg K)]
$20 < T_{st} < 600$	$425 + 0.773 \cdot T_{st} - 1.69 \times 10^{-3} \cdot T_{st}^{2} + 2.22 \times 10^{-6} \cdot T_{st}^{3}$
$600 < T_{st} < 735$	$666 + 13002/(738 - T_{st})$
$735 < T_{st} < 900$	$545 + 17820/(T_{st} - 731)$
$900 < T_{st} < 1200$	650

Table 13.3 Summary of thermal properties of carbon steel including derived volumetric specific enthalpy according to Eurocode 3, EN 1993-1-2

Temp	k_{st}	ρ_{st}	c_{st}	e_{st}	e_{st}
[°C]	[W/(m K)]	[kg/m^3]	[J/(kg K)]	[J/(m^3 K)]	[Wh/(m^3 K)]
0	54	7850	425	0	0
100	51	7850	488	0.360E+09	99,870
200	47	7850	530	0.760E+09	211,000
300	44	7850	565	1.19E+09	330,300
400	41	7850	606	1.65E+09	457,800
500	37	7850	667	2.15E+09	596,100
600	34	7850	760	2.70E+09	751,100
700	31	7850	1008	3.37E+09	934,300
735	30	7850	5000	4.20E+09	1,091,000
800	27	7850	803	5.03E+09	1,309,000
900	27	7850	650	5.58E+09	1,464,000
1200	27	7850	650	7.12E+09	1,890,000

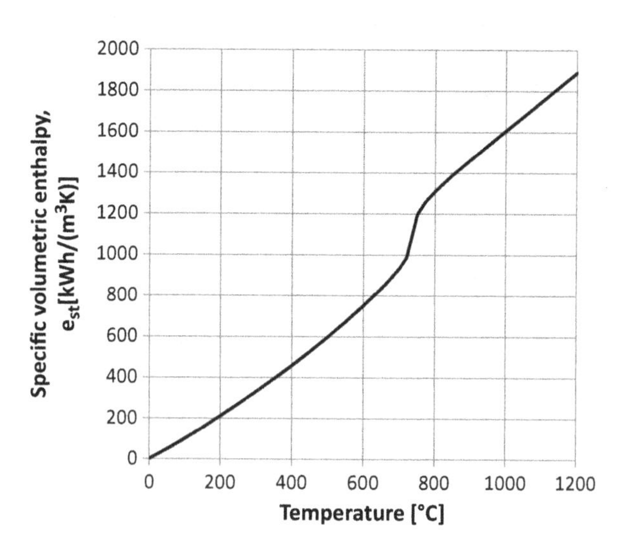

Fig. 13.3 Volumetric enthalpy of carbon steel according to Table 13.3 derived from Eurocode 3, EN 1993-1-2

Table 13.4 Thermal conductivity and specific heat capacity of stainless steel vs. temperature according to Eurocode 3, EN 1993-1-2

Temperature [°C]	Conductivity [W/(m K)]	Specific heat capacity [J/(kg K)]
$20 < T_{st} < 1200$	$14.6 + 0.0127 \cdot T_{st}$	$450 + 0.280 \cdot T_{st} - 0.291 \cdot 10^{-3} \cdot T_{st}^{2} + 0.134 \cdot 10^{-6} \cdot T_{st}^{3}$

Table 13.5 Dimensions of hot-rolled HEB steel sections according to EN 10025-1. The last column corresponds to A_{st}

HEB	Height [mm]	Width [mm]	Web thickness [mm]	Flange thickness [mm]	Weight [kg/m]	Surface area [m²/m]
100	100	100	6	10	20.8	0.567
120	120	120	6.5	11	27.2	0.686
140	140	140	7	12	34.4	0.805
160	160	160	8	13	43.4	0.918
180	180	180	8.5	14	52.2	104
200	200	200	9	15	62.5	1.15
220	220	220	9.5	16	72.8	1.27
240	240	240	10	17	84.8	1.38
260	260	260	10	17.5	94.8	1.5
280	280	280	10.5	18	105	1.62
300	300	300	11	19	119	1.73
320	320	300	11.5	20.5	129	1.77
340	340	300	12	21.5	137	1.81
360	360	300	12.5	22.5	145	1.85
400	400	300	13.5	24	158	1.93
450	450	300	14	26	174	2.03
500	500	300	14.5	28	191	2.12
550	550	300	15	29	203	2.22
600	600	300	15.5	30	216	2.32
650	650	300	16	31	229	2.42
700	700	300	17	32	245	2.52
800	800	300	17.5	33	267	2.71
900	900	300	18.5	35	297	2.91
1000	1000	300	19	36	320	3.11

13.2 Example of Hot-Rolled Steel Section Dimensions

Dimensions of hot-rolled steel sections can be found for instance in suppliers catalogues or on the internet web. As an example dimensions of HEB wide-flange steel I-sections according to the European standard EN 10025-1 are given in Table 13.5.

13.3 Protected Steel Sections Assuming Lumped-Heat-Capacity

The assumption of *lumped heat* or *uniform steel temperature* as often done in fire protection engineering calculations (see, e.g. Eurocode 3) is in particular a reasonable approximation when calculating temperature of protected steel sections exposed to fire on all four sides. The assumption of uniform heat implies that the heat conductivity is assumed infinite and the thermal mass is concentrated, lumped, to one point, see Sects. 3.1 and 7.1.

Then in addition *the fire* and *the exposed surface* temperatures are assumed equal which implies that the heat transfer resistance between the fire gases and the protection surface is negligible. That means the inverse of the total heat transfer coefficient by radiation and convection is assumed negligible in comparison with the heat resistance of the insulation R_k, i.e. the thickness over the conductivity d_{in}/k_{in} of the insulation, cf. Fig. 3.3. This is an accurate approximation as the radiation heat transfer coefficient is very high at elevated fire temperatures. It facilitates calculations and it is on the safe side as it overestimates steel temperatures.

The heat transfer to the steel may then be calculated as

$$\dot{q}_{tot} = A_{st}\left(\frac{k_{in}}{d_{in}}\right)(T_f - T_{st}) \tag{13.2}$$

where A_{st} is the fire-exposed area per unit length, T_f and T_{st} are the fire and steel temperatures, respectively. If in addition the heat capacity of the insulation is negligible in comparison to that of the steel, the transient heat balance of the steel section becomes,

$$A_{st}\left(\frac{k_{in}}{d_{in}}\right)(T_f - T_{st}) = c_{st}\rho_{st}V_{st}\frac{\partial T_{st}}{\partial t} \tag{13.3}$$

That is the heat entering the steel section is equal to the heat stored per unit time proportional to the rate of temperature rise. c_{st} and ρ_{st} are the specific heat capacity and density, respectively, of steel and V_{st} the volume per unit length of the steel section. When estimating the conductivity of the insulation the temperature of the insulation may be assumed as the mean of the fire and the steel temperatures.

In cases of *heavy insulations* when the heat capacity of the insulation need be considered a more rigorous analysis is required as shown in Sect. 13.3.1.

From Eq. 13.3 the forward difference scheme

$$T_{st}^{i+1} = T_{st}^{i} + \left(\frac{A_{st}}{V_{st}}\right)\frac{\Delta t}{\rho_{st}\cdot {c_{st}}^{i}}\left(\frac{k_{in}}{d_{in}}\right)\left(T_f^{i+1} - T_{st}^{i}\right) \tag{13.4}$$

where Δt is a chosen time increment. The specific heat ${c_{st}}^{i}$ is taken at the temperature level T_{st}^{i} (if assumed varying with temperature).

The relation A_{st}/V_{st} is denoted the *section factor* or *shape factor*. It has the dimension one over length [m^{-1}]. The shape factor can be replaced by its reciprocal, the *effective thickness* of the steel $\overline{d_{st}}$ identified as

$$\overline{d_{st}} = \frac{V_{st}}{A_{st}} \tag{13.5}$$

Instructions on how to obtain shape factors for various steel sections are given in Table 13.6 taken from Eurocode 3 [3]. The area A_{st} of contour encasements such as spray fire protection material is generally taken as the perimeter of the section times the unit length. For board protections forming hollow encasements, the perimeter may be assumed as the boxed value as shown in the second row of Table 13.6. Even if there is a clearance around the member, the same boxed value may be applied. For steel sections fire exposed on three sides the perimeter is reduced accordingly as shown in the third and fourth rows of Table 13.6. Thus the interface between the steel and, for example, a concrete slab is treated as an adiabatic surface and hence the cooling effects of the steel section is ignored. Therefore this crude approximation model yields considerably higher temperature than it could be expected in reality. To accurately incorporate the cooling effects 2D finite element calculations are required.

Alternatively the steel section volume per unit length V_{st} may be obtained as the *weight per unit length* m_{st} (often tabulated in catalogues of steel providers) over the steel density ρ_{st}, i.e.

$$V_{st} = \frac{m_{st}}{\rho_{st}} \tag{13.6}$$

Analytical solutions can be derived only when *constant* conductivity of the protection material and specific heat of the steel are assumed. If in addition the fire temperature is assumed to suddenly rise to constant temperature, can the steel temperature be obtained as shown in Sect. 3.1.2 as

$$\frac{T_{st} - T_i}{T_f - T_i} = 1 - e^{-\frac{t}{\tau}} \tag{13.7}$$

where τ is identified as a time constant which for a protected steel section becomes

$$\tau = \left(\frac{V_{st}}{A_{st}}\right)\rho_{st}c_{st}\left(\frac{d_{in}}{k_{in}}\right) = \overline{d_{st}}\rho_{st}c_{st}\left(\frac{d_{in}}{k_{in}}\right) \tag{13.8}$$

In some special cases with varying fire temperatures the steel temperatures may be calculated analytically as shown in Sect. 13.3.2 where efficient and compact diagrams which facilitates estimations of steel temperatures are shown.

In general, however, the time constant τ cannot be assumed constant as the thermal properties of the insulation as well as of the steel vary with temperature and

Table 13.6 Section factor A_{st}/V_{st} for steel members insulated by fire protection material. From Eurocode 3 [3]

Sketch	Description	Section factor (A_{st}/V_{st})
	Contour encasement of uniform thickness	$\frac{\text{steel perimeter}}{\text{steel cross-section area}}$
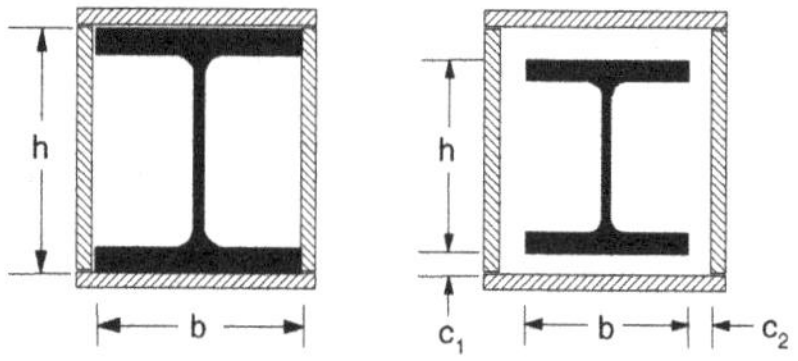	Hollow encasement of uniform thickness[a]	$\frac{2(b+h)}{\text{steel cross-section area}}$
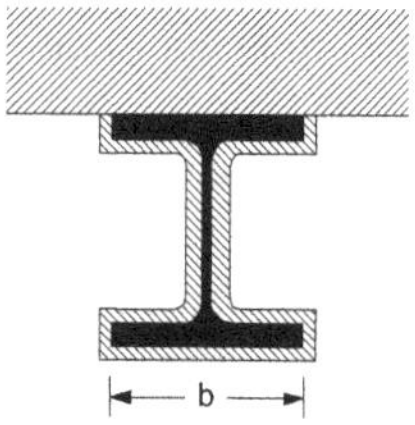	Contour encasement of uniform thickness, exposed to fire on three sides	$\frac{\text{steel perimeter } 2h+b}{\text{steel cross-section area}}$
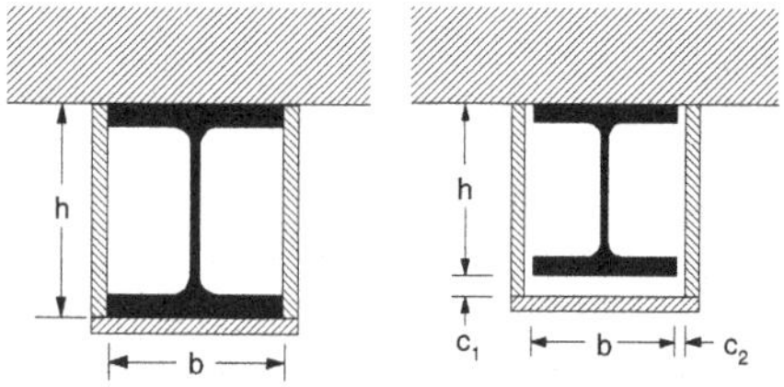	Hollow encasement of uniform thickness, exposed to fire on three sides[a]	$\frac{2\ h+b}{\text{steel cross-section area}}$

The clearance dimensions c_1 and c_2 should not normally exceed $h/4$

time, and as the fire temperature T_f generally varies with time. As an alternative to Eq. 13.4 the steel temperature can be calculated by forward difference recursion formula

$$T_{st}^{i+1} = \frac{\Delta t}{\tau^i} T_f^{i+1} + \left(1 - \frac{\Delta t}{\tau^i}\right) T_{st}^i \tag{13.9}$$

where Δt is a chosen time increment. The suffixes denote the numerical order of the time increments. When the thermal properties vary with temperature, the time constant τ need be updated at each time increment.

The forward difference scheme of Eq. 13.12 is numerically stable if the time increment is less than the time constant at each time increment i, i.e.

$$\Delta t \le \tau_i \tag{13.10}$$

In practice time increments Δt longer than 10 % of the time constant should not be used to assure numerical stability and accuracy. When choosing the time increment it is also necessary to make it short enough to be able to follow the thermal exposure changes with time.

The recursion formulas according to Eq. 13.9 are preferably solved with a spreadsheet program such as MS-Excel. For clarification examples are shown below on how the formula is used.

Example 13.1 A steel column with a section factor 200 m^{-1} is protected with a 25 mm non-combustible board with a conductivity of 0.1 W/(m K). The column is exposed to fire and the exposed insulation surface suddenly reaches a temperature of 1000 °C. Assume constant thermal properties and uniform steel temperature (lumped heat). The density and specific heat of steel are assumed to be 7850 kg/m^3 and 460 W s/(kg K), respectively. The initial temperature $T_i = 20$ °C. Calculate the steel temperature after

(a) 9 min using the analytical exact solution according to Eq. 13.7
(b) 60 min using the analytical exact solution according to Eq. 13.7
(c) 9 min using the numerical solution according to Eq. 13.9 and compare with (a)

Solution According to Eq. 13.8 the time constant $\tau = \frac{460 \cdot 7850 \cdot 0.025/0.1}{200} = 4514\,\text{s}$.

(a) After 9 min $\frac{t}{\tau} = 9 \cdot 60/4514 = 0.12$ and according to Eq. 13.7 or Fig. 3.4 $\left(1 - e^{-\frac{t}{\tau}}\right) = 0.113$ and the steel temperature $T_{st} = 20 + (1000 - 20)*0.113 = 131\,°\text{C}$.
(b) After 60 min $\frac{t}{\tau} = 60 \cdot 60/4514 = 0.80$ and the steel temperature $T_{st} = 20 + (1000 - 20)*0.55 = 560$ °C.
(c) Assume a time increment $\Delta t = 3\,\text{min} = 180\,\text{s}$. Then according to Eq. 13.9 at t = 180 s $T_{st}^1 = \left(\frac{180}{4514}\right) \cdot 1000 + \left(1 - \frac{180}{4514}\right) \cdot 20 = 59\,°\text{C}$, at t = 360 s $T_{st}^2 = \left(\frac{180}{4514}\right) \cdot 1000 + \left(1 - \frac{180}{4514}\right) \cdot 59 = 96\,°\text{C}$ and at 540 s $T_{st}^2 = \left(\frac{180}{4514}\right) \cdot 1000 + \left(1 - \frac{180}{4514}\right) \cdot 96 = 132\,°\text{C}$. Notice that the numerical solution (b) is only 1 °C more than the exact solution according to (a).

Example 13.2 The same column as in Example 13.2 is exposed to a standard fire time–temperature curve according to ISO 834. Calculate the steel temperature after 9 min.

Solution Apply the recursion formula according to Eq. 13.9. Choose a time increment $\Delta t = 180\,\text{s}$ ($\ll$10 % of τ). The temperature at 3, 6 and 9 min is according to Eq. 12.1, 7.7, 10.17 and 12.5, respectively. Then according to Eq. 13.9 at t = 180 s $T_{st}^1 = \left(\frac{180}{4514}\right) \cdot 228 + \left(1 - \frac{180}{4514}\right) \cdot 20 = 28.3\,°\text{C}$, at t = 360 s $T_{st}^2 = \left(\frac{180}{4514}\right) \cdot 312 + \left(1 - \frac{180}{4514}\right) \cdot 28.3 = 36.6\,°\text{C}$ and at 540 s $T_{st}^2 = \left(\frac{180}{4514}\right) \cdot 365 + \left(1 - \frac{180}{4514}\right) \cdot 36.6 = 49.7\,°\text{C}$.

13.3.1 Protection with Heavy Materials

The heat capacity of fire protections has normally an insignificant influence on the steel temperature rise rate. However, it will considerably reduce the steel temperature rise of sections protected with relatively heavy protections materials. The protection will then add to the heat capacity of the system and it will cause a delay in the temperature rise of the steel section. A simple approximate approach is then to lump a third of the heat capacity of the insulation to the steel section heat capacity and to add a term considering the time delay [42–44]. Eq. 13.4 may then be modified the more general formulation as

$$T_{st}^{j+1} = T_{st}^{j} + \Delta t \frac{\left(T_f^{j+1} - T_{st}^{j}\right)}{\tau\left(1 + \frac{\mu}{3}\right)} + \left(e^{\frac{\mu}{\eta}} - 1\right)\left(T_f^{j+1} - T_f^{j}\right) \tag{13.11}$$

where τ is as specified in Eq. 13.7 and μ is the ratio between the heat capacity of the insulation and the steel,

$$\mu = \frac{A_{st} \cdot d_{in} \cdot \rho_{in} \cdot c_{in}}{V_{st} \cdot \rho_{st} \cdot c_{st}} \tag{13.12}$$

ρ_{in} and c_{in} are the density and specific heat capacity of the protection material, respectively. The latter term of Eq. 13.11 represents a time delay due to the heat capacity of the protection. $(T_f^{j+1} - T_f^{j})$ is the fire temperature rise between two time increments. Notice that when the heat capacity of the protection is much smaller than that of the steel, μ vanishes and Eq. 13.11 becomes identical to Eq. 13.4.

The value of the parameter η in the last term of Eq. 13.11 was obtained by comparisons with accurate finite element calculations. For steel sections exposed to the ISO/EN standard time–temperature curve accurate approximations are obtained by choosing $\eta = 5$, see Example 13.3 and Fig. 13.4.

For fire temperatures assumed to instantaneously rise to a given temperature, $\eta = 10$ yields very similar steel temperatures in comparison to accurately calculated temperatures. This value has been adopted by Eurocode 3, EN 1993-1-2 [3]. It yields higher steel temperatures than choosing the more accurate value $\eta = 5$.

Example 13.3 A steel section with a section factor of 200 m^{-1} and an initial temperature of 20 °C is exposed to a standard fire curve according to ISO 834. It is protected with 20-mm-thick high density material assumed to have the same properties as concrete. Assume material properties as given in Table 1.2. Calculate the steel temperature development.

Solution According to Eq. 13.7 $\tau = \frac{\rho_{st} \cdot c_{st}}{(A_{st}/V_{st})}\left(\frac{d_{in}}{k_{in}}\right) = \frac{7850 \cdot 460 \cdot 0.02}{200 \cdot 1.7} = 212\,\text{s}$ and Eq. 9.12 $\mu = \frac{A_{st} \cdot d_{in} \cdot \rho_{in} \cdot c_{in}}{V_{st} \cdot \rho_{st} \cdot c_{st}} = \frac{200 \cdot 0.02 \cdot 2300 \cdot 900}{7850 \cdot 460} = 2.29$. Then the recursion formula Eq. 13.11 may be applied. Steel temperatures obtained by an MS-Excel application are given in

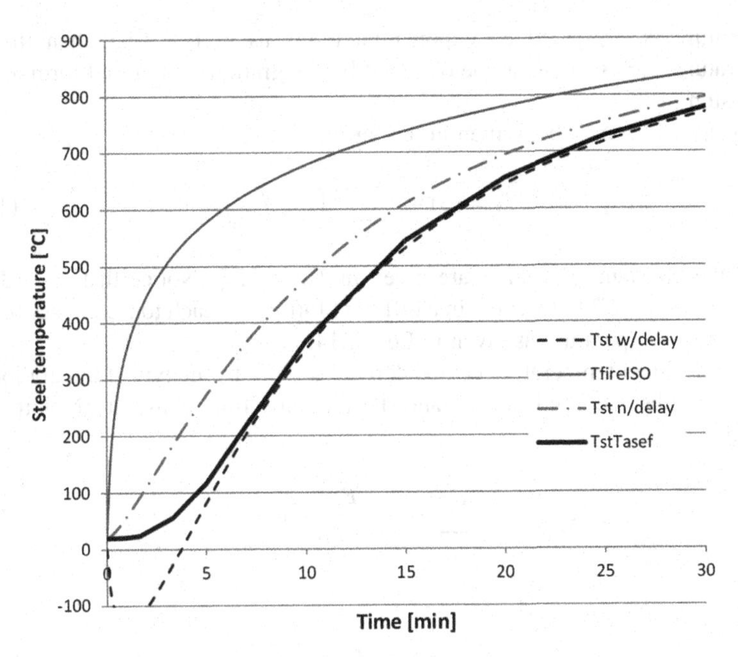

Fig. 13.4 Example of steel temperature of a steel section protected by a heavy protection material exposed to the ISO 834 standard time–temperature fire curve. Temperature calculated with the delay and with no delay according to the third term according to Eq. 13.11. For comparison the steel temperature as calculated accurately with the finite element code TASEF

Fig. 13.4. For comparison accurately finite element calculated temperatures are also shown. Notice how well the temperature calculated according to Eq. 13.11 matches the accurate solution except for the 5 min when the temperature goes down even below zero. In addition the temperatures are shown which are calculated without considering the delay expressed by the parameter μ larger than zero in the third term on the right-hand side of Eq. 13.11.

13.3.2 Protected Steel Sections Exposed to Parametric Fire Curves

As described in Sect. 12.2 the concept of parametric fires has been introduced in Eurocode 1 [35] as a convenient way of expressing a set of post-flashover design fires.

When using parametric design fires the temperature of protected steel sections can of course be obtained by numerical calculations according to Eq. 13.9. Then non-linear phenomena such as temperature-dependent material properties may be considered. However, if the thermal properties are assumed constant and the fire temperature is expressed by exponential terms as in Eq. 12.5, then the steel temperature rise vs. time can be obtained by integration as a closed form analytic expression [45].

Equation 12.5 may be written in the form

$$T_f = 20 + \sum_{i=0}^{3A} \left(B_i e^{-\beta_i t^*} \right) \tag{13.13}$$

where the constants B_i and β_i are given in Table 13.7. Notice that Eq. 13.13 is identical to Eq. 12.5 but written in a different format to reach to a compact solution for the steel temperature as given in Eq. 13.14.

Then the steel temperature can be derived exactly by analytical integration as a function of the modified time t^* and the modified time constant τ^* of the steel section as

$$T_{st} = 20 + \sum_{i=0}^{3} \left[\frac{B_i}{1 + \beta_i \tau^*} \left(e^{\beta_i t^*} - e^{-\frac{t^*}{\tau^*}} \right) \right] \tag{13.14}$$

where

$$\tau^* = \Gamma \cdot \tau \tag{13.15}$$

The protected steel section time constant τ is given in Eq. 13.8. The relation between the temperature rise vs. modified time as expressed in Eq. 13.14 is also given in the diagram shown in Fig. 13.5a, b for various modified time constants τ^*. The two diagrams are the same but with different time and temperature scales. Notice that Eq. 13.14 and Fig. 13.5 may be used for the ISO 834 standard fire exposures assuming $\Gamma = 1$ as the parametric fire curve in the heating phase then is very close to the standard curve, see Fig. 12.3.

The use of parametric fire curves on insulated steel sections is demonstrated below.

Example 13.4 Consider a steel section with a shape factor = 200 m^{-1} with a 25-mm-thick protection board having a constant thermal conductivity of 0.1 W/(m K). The steel density and specific heat capacity are 7850 kg/m^3 and 460 J/(kg K), respectively. The section time constant may then be obtained from Eq. 13.15 as $\tau = 4514$ s = 75 min = 1.25 h. Then if the section is exposed to ISO 834 standard fire ($\Gamma = 1$) for 60 min, a temperature of 462 °C may be obtained from Eq. 13.14 or from Fig. 13.5. If the same section is exposed to a more slowly growing fire with a $\Gamma = 0.5$, then $\tau^* = \Gamma \cdot \tau = 37.5$ min and the temperature after 60 min may be found for a modified time of $t^* = \Gamma \cdot t = 30$ min to be 405 °C. On the other hand, if the section is exposed to a fast growing fire with $\Gamma = 3.0$, then $\tau^* = 3.0 \cdot 75 = 225$ min and $t^* = 3.0 \cdot 60 = 180$ min, and the steel temperature can be obtained from Eq. 13.13 or from Fig. 13.5 as 552 °C. Notice that the maximum steel temperature for a given fire exposure time increases considerably with an increasing Γ-factor. It

Table 13.7 Constants in the analytical expression of the parametric fire curve

Term number, i	0	1	2	3
B_i (°C)	1325	−430	−270	−625
β_i (h^{-1})	0	−0.2	−1.7	−19

must, however, also be kept in mind that the fire duration for a given fuel load is proportional to the inverse of the opening factor included in the Γ-factor.

The diagrams of Figure 13.6a, b show the temperature development of a steel structure with the same dimensions and protection as described above. The two cases are assuming the same fire $q_f = 200\,\mathrm{kJ/m^2}$ and thermal inertia of the surrounding structure $\sqrt{k \cdot \rho \cdot c} = 1160\,\mathrm{W\,s^{1/2}/(K\,m^2)}$. According to Eq. 12.8 fire duration can be calculated to be $t_d = 60$ min and 120 min $(\chi = 0.2{*}10^{-3}\,[\mathrm{hm^{3/2}/MJ}])$, with $\Gamma = 1$ and 0.25, respectively. Notice that the steel temperature reaches its maximum when it is equal to the cooling phase fire temperature. According to the diagram of Figure 13.6a the steel temperature is 450 °C at $t_d = 60$ min and reaches its maximum 570 °C after 105 min. The corresponding temperatures according to diagram in Figure 13.6b are 585 °C after 120 min and the maximum steel temperature is 670 °C after 180 min. Thus the steel section reaches a higher temperature for the lower opening factor, given the fire load and the thermal properties of the surrounding compartment boundaries remains the same.

For more detailed information on how to apply parametric fire curves according to standard, see Eurocode 1 [35].

13.4 Unprotected Steel Sections

The temperature of unprotected, uninsulated or bare steel sections depends on the fire temperature and very much on the heat transfer conditions between fire gases and steel surfaces. It is a boundary condition of the 3rd kind, see Sect. 1.1.3, where the only thermal resistance between the fire and the steel is due to the heat transfer conditions which therefore becomes decisive for the steel temperature development. The boundary condition is highly non-linear as it varies very much with temperature due to radiation. The same type of compact formula and diagrams as for insulated steel sections can therefore not be developed.

The total heat flux by radiation and convection $\dot{q}''_{tot}$ is given in Eq. 4.17 or Eqs. 4.18 and 4.19. Steel temperatures can then be obtained from differential heat balance equations in a similar way as for protected steel sections (cf. Eqs. 13.2 and 13.3).

According to Eq. 4.17 and the procedures as outlined in Sect. 7.1 the heat flux by radiation and convection can be written as

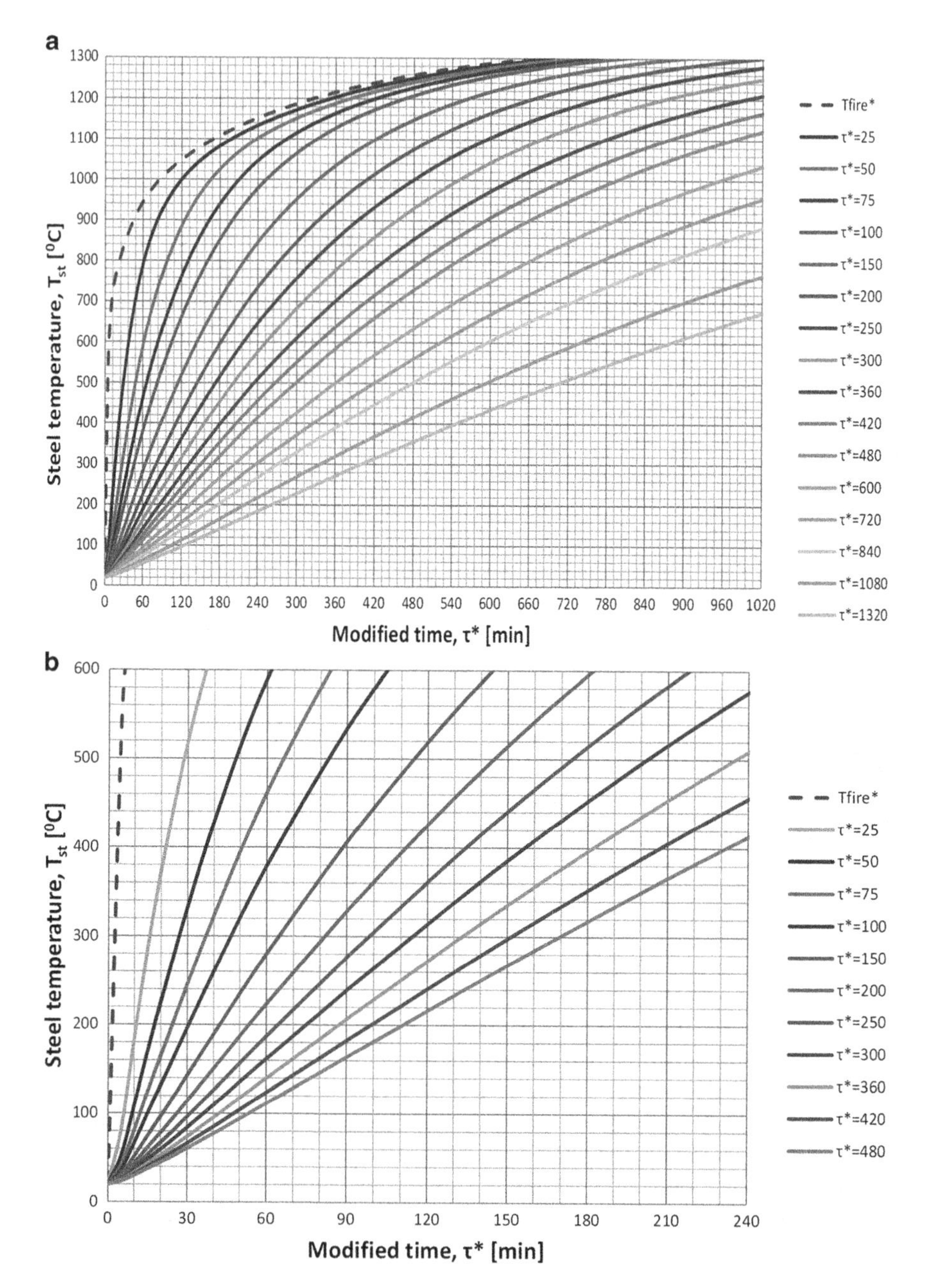

Fig. 13.5 Temperature of various protected steel sections exposed to parametric fires in the heating phase vs. modified time t^*. The thermal properties of the steel sections are embedded in the modified time constants τ^*, see Eq. 13.15. The *bottom diagram* is a magnification of the top

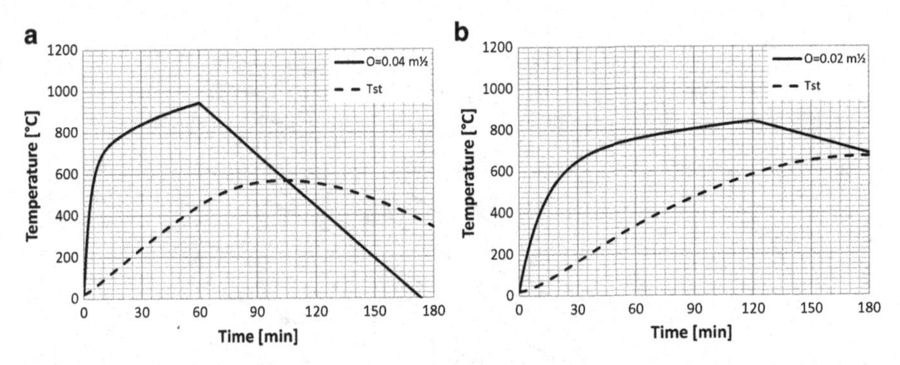

Fig. 13.6 Fire and steel temperatures calculated numerically according to Eq. 13.4 based on parametric fire curves with $q_f = 200\,\text{kJ/m}^2$, $\sqrt{k\rho c} = 1160\,\text{W}\,\text{s}^{1/2}/(\text{K}\,\text{m}^2)$ and $O = 0.04$ m^{½} (diagram a) and $O = 0.02$ m^{½} (diagram b), respectively. See Example 13.4. (**a**) $O = 0.04$ m^{½}, $\Gamma = 1$ and $t_d = 60$ min (**b**) $O = 0.02$ m^{½}, $\Gamma = 0.25$ and $t_d = 120$ min

$$\dot{q}''_{tot} = \varepsilon_{st} \cdot \sigma\left(T_f^4 - T_s^4\right) + h_c\left(T_f - T_s\right) \tag{13.16}$$

when the radiation and convection temperatures are assumed equal, i.e. $T_r = T_g = T_f$. This heat flux shall balance with the heat stored in the steel section, i.e.

$$A_{st} \cdot \dot{q}''_{tot} = c_{st} \cdot \rho_{st} \cdot V_{st} \frac{\partial T_{st}}{\partial t} \tag{13.17}$$

and the steel temperature can then be obtained by the numerical time integration scheme

$$T^{j+1} = T^j + \frac{A_{st}}{c_{st} \cdot \rho_{st} \cdot V_{st}} \left[\varepsilon_{st} \cdot \sigma\left(T_f^{j\,4} - T^{j\,4}\right) + h_c\left(T_f^j - T^j\right)\right] \cdot t \tag{13.18}$$

where Δt is the time increment and the superscript j the time increment number. The heat capacity of the steel may be updated at each time step to consider changes dependent on temperature.

Figure 13.7 shows steel temperature developments of steel sections with various section factors assuming constant values of c_{st}, ρ_{st}, ε_{st} and h_c.

Equation 13.20 is a forward difference scheme which is numerically stable and accurate only for limited values of the time increment. The stability criterion for the explicit numerical scheme may be expressed as

$$\Delta t^i \leq \tau^i = \left\{\frac{V_{st}\rho_{st}c_{st}}{A_{st}h_{tot}}\right\}^i \tag{13.19}$$

where h_{tot} is the total heat transfer coefficient

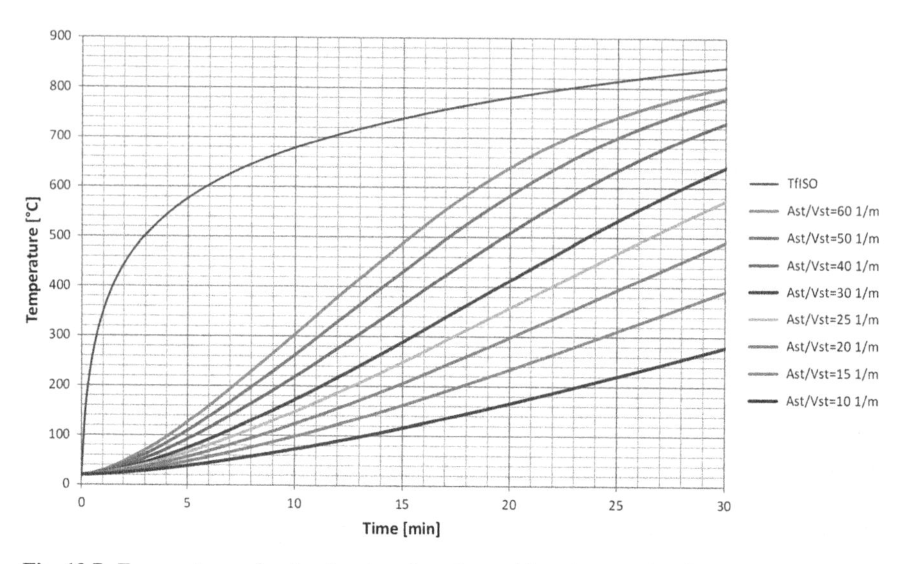

Fig. 13.7 Temperatures of *uninsulated* steel sections with various section factors exposed to the standard ISO 834 time–temperature curve calculated according to Eq. 13.18 with $c_{st} = 460\,\mathrm{J/(kg\,K)}$, $\rho_{st} = 7850\,\mathrm{kg/m^3}$, $\varepsilon_{st} = 0.7$ and $h_c = 25\,\mathrm{W/(m^2\,K)}$

$$h_{tot} = h_r + h_c = \varepsilon_{st} \cdot \sigma\left(T_f^2 + T_{st}^2\right)\left(T_f + T_{st}\right) + h_c \tag{13.20}$$

As the total heat transfer coefficient h_{tot} will increase substantially with the temperature level, c.f. Eq. 13.20, the time constant and critical time step will decrease accordingly.

In practice it is recommended to keep the time increments less than 10 % of the current time constant, i.e.

$$\Delta t^i \leq 0.1 \left\{\frac{V_{st}\rho_{st}c_{st}}{A_{st}h_{tot}}\right\}^i \tag{13.21}$$

Principles for calculating the section factors according to Eurocode 3 [3] for various types of configurations of unprotected steel members can be found in Table 13.8.

As well as for protected steel sections the volume V_{st} may be calculated as the weight per unit length m_{st} over the steel density ρ_{st}, see Eq. 13.6. The weight per unit length m_{st} of steel sections is often tabulated in catalogues of steel supplier.

Table 13.8 Section factor A_{st}/V_{st} for unprotected steel members

Open section exposed to fire on all sides: $\frac{A_{st}}{V_{st}} = \frac{\text{perimeter}}{\text{cross-section area}}$ 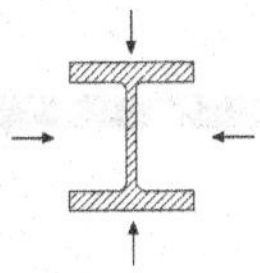 (note shadow effects Sect. 13.4.1)	Tube exposed to fire all around: $A_{st}/V_{st} = 1/t$ 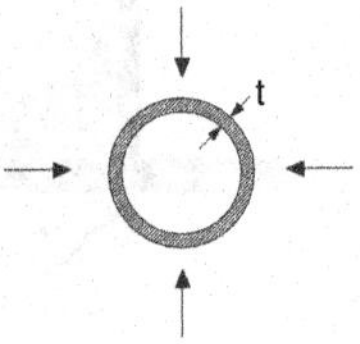
Open section exposed to fire on three sides: $\frac{A_{st}}{V_{st}} = \frac{\textit{surface exposed to fire}}{\textit{cross-section area}}$ 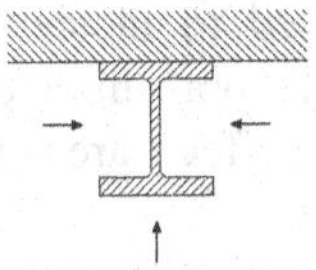 (note shadow effects Sect. 13.4.1)	Hollow section (or welded box section of uniform thickness) exposed to fire on all sides: If $t \ll b$: $A_{st}/V_{st} \approx 1/t$ 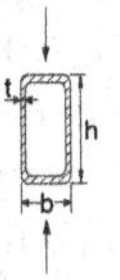
I-section flange exposed to fire on three sides: $A_{st}/V_{st} = (b + 2t_f)/(b \cdot t_f)$ If $t \ll b$: $A_{st}/V_{st} \approx 1/t_f$ 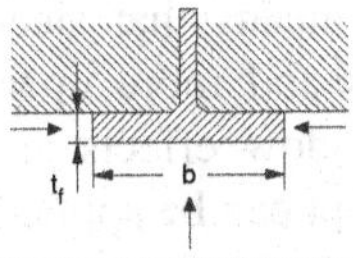	Welded box section exposed to fire on all sides: $\frac{A_{st}}{V_{st}} = \frac{2(b+h)}{\textit{cross-section area}}$ If $t \ll b$: $A_{st}/V_{st} \approx 1/t$ 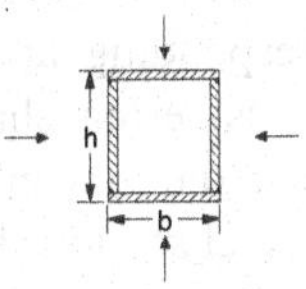
Angle exposed to fire on all sides: $A_{st}/V_{st} = 2/t$ 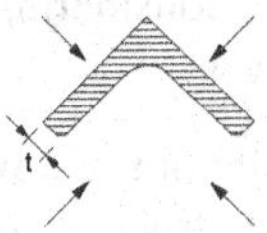 (note shadow effects Sect. 13.4.1)	I-section with box reinforcement, exposed to fire on all sides: $\frac{A_{st}}{V_{st}} = \frac{2(b+h)}{\text{cross-section area}}$ 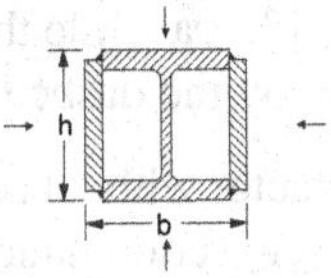
Flat bar exposed to fire on all sides: $A_{st}/V_{st} = 2(b+t)/(b \cdot t)$ If $t \ll b$: $A_{st}/V_{st} \approx 2/t$ 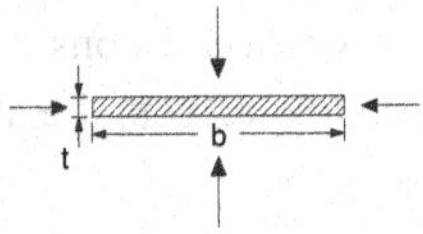	Flat bar exposed to fire on three sides: $A_{st}/V_{st} = (b+2t)/(b \cdot t)$ If $t \ll b$: $A_{st}/V_{st} \approx 1/t$ 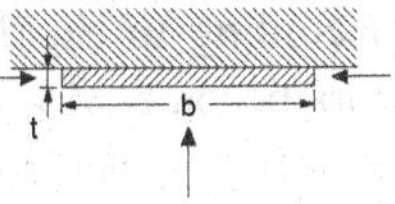

From Eurocode 3 [3]

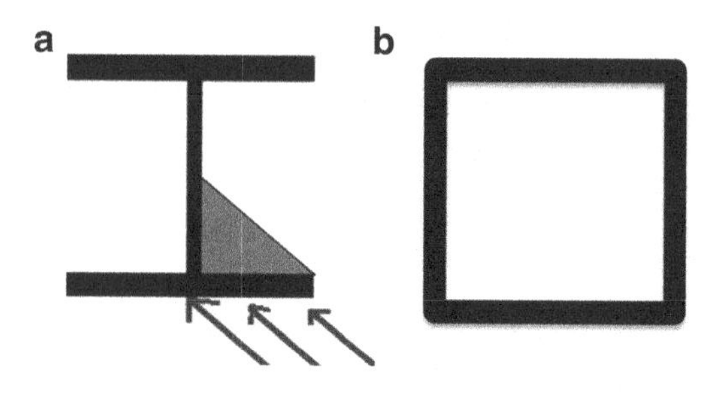

Fig. 13.8 Shadow effects are applicable to open sections where surfaces are partly shadowed against incident radiation. (**a**) Open section (**b**) Closed section

13.4.1 Shadow Effects

When an open section such as an I-section is exposed to fire, the heat transfer by radiation will be partly shadowed as indicated by Fig. 13.8a, see Eurocode 3 [3] and [46]. The surfaces between the two flanges are then not exposed to incident radiation from the surrounding fire from the full half-sphere but only from a limited angle, i.e. the incident radiation to these surfaces is reduced. Shadow effects are not applicable to closed sections such as tubes as shown in Fig. 13.8b.

As a matter of fact a section will only receive as much heat by radiation from the fire as if it had the same periphery as a "boxed" section, see Fig. 13.9a, b. Therefore the area per unit length A_{st} may be replaced by the so-called boxed area A_b in the Eq. 13.18. This will reduce the influence by convection heat transfer but as the radiation heat transfer mode dominates at elevated temperature this approximation may be accepted although it is non-conservative. The boxed area A_b is typically for an I-section 30 % less than the corresponding area A_{st}. This means that steel temperature will be reduced when considering shadow effects and more open steel sections can be accepted without thermal protection. Shadow effects are particularly important for unprotected steel sections but the concept can be applied to other types of structures as well.

The surface area of an I-beam attached to a concrete slab or wall may be reduced in a similar way as indicated in Fig. 13.10. According to Table 13.8 the surface area A_{st} can be calculated as shown in Fig. 13.10a while the reduced area A_b considering shadow effects is calculated as shown by the dashed line in Fig. 13.10b.

Example 13.5 Calculate the section factor without and with considering of shadow effects of an unprotected HE300B steel section attached to a concrete structure as shown in Fig. 13.10.

Solution Dimensions of an HE300B section can be found in Table 13.5. Thus $A_{st} = 2H + 3W - 2t_w = 2 \cdot 300 + 3 \cdot 300 - 2 \cdot 11 = 1478\,\text{mm}$ and $A_b = 2H + W = 2 \cdot 300 + 300 = 900\,\text{mm}$. The section weight $m_{st} = 119$ kg per unit length according to Table 13.5. Thus according to Eq. 13.6 $V_{st} = m_{st}/\rho_{st} = 119/7850 = 0.0151\,\text{m}^2$ and the section factors becomes $A_{st}/V_{st} = 98\,\text{m}^{-1}$ when not considering shadow effects and $A_b/V_{st} = 60\,\text{m}^{-1}$ when considering shadow effects, i.e. a reduction of about 40 %.

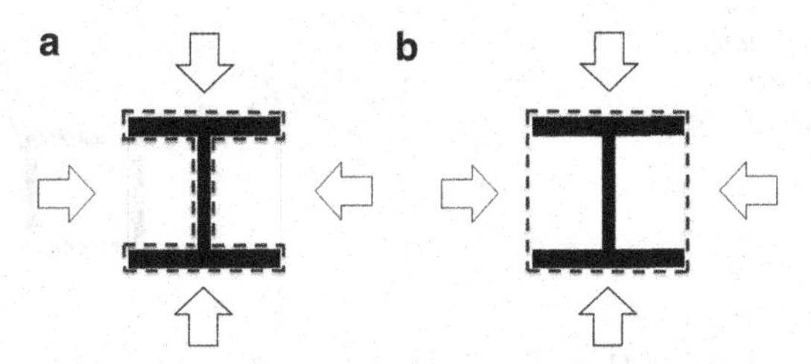

Fig. 13.9 Illustration of the *shadow effect* of I-section exposed to fire from four sides. (**a**) Area without considering shadow effects, A_{st} (**b**) The boxed area considering shadow effects, A_b

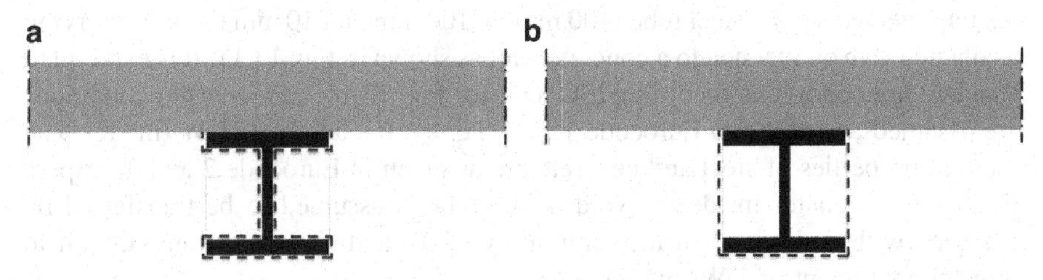

Fig. 13.10 Periphery considering and not considering shadow effects for steel profiles attached to concrete structures. (**a**) Periphery according to Table 13.8, A_{st} (**b**) Periphery considering shadow effects, A_b

Comment: Still such a section would get a temperature of 500 °C already after 15 min according to Fig. 13.7. The results are, however, conservative as the heat transferred from the steel to the concrete is not considered in this type of calculations. Temperatures calculated with the finite element code TASEF including shadow effects as well as effects of cooling to the concrete structure are shown in Sect. 13.5.3.

13.5 Examples of Steel Temperatures Calculated Using a Finite Element Code

The steel section temperature analyses above assume uniform steel temperatures or lumped heat. This is often a very crude approximation. It leads indeed in general to solutions on the safe side, i.e. the temperatures are overestimated, but often to overdesign and thereby to unnecessary costs. Unsafe conditions may, however, occur in sections where parts such as webs are considerably thinner than the flanges.

For more precise analyses numerical calculations are needed employing, e.g. finite element computer codes. Some examples are shown in the sections below.

Fig. 13.11 A bare *square* steel tube section carrying a concrete slab or attached to a concrete wall

13.5.1 Unprotected Square Steel Tube Section Attached to a Concrete Slab or Wall

An unprotected square steel tube (100 mm by 100 mm and 10 mm thick) is carrying a concrete slab or attached to a concrete wall as shown in Fig. 13.11. It is exposed to standard fire conditions according ISO 834, see Fig. 13.11. Heat transfer conditions are assumed according to Eurocode 1 [35], i.e. $\varepsilon = 0.8$ and $h = 25$ W/(m^2 K). The thermal properties of steel and concrete are as given in Eurocode 2 and 3, respectively. Heat transfer inside the void of the tube is assumed to be transferred by radiation with an internal surface emissivity of 0.8 and by convection with a heat transfer coefficient of 1 W/(m^2 K).

The temperature calculation was carried with the finite element computer code TASEF [19]. The finite element discretization model including element node numbers is shown in Fig. 13.12a. Calculated steel temperatures vs. time are shown in Fig. 13.12b, the bottom flange (node 1) and two of the top flange (nodes 5 and 35). Notice that the temperature of the bottom flange is considerably higher than that of the top flange. The difference decreases, however, in the end of the exposure as the radiation heat transfer between the flanges becomes more efficient at higher temperature levels and the concrete slab is heated. The heat transfer in the void levels out the temperature as heat is transferred between surfaces, it cools the exposed flange and heats the flange attached to the concrete. Figure 13.12c shows a temperature contour after 15 min.

13.5.2 Encased I-Section Connected to a Concrete Structure

An HE300B steel section attached to a concrete structure, wall or slab, is protected by gypsum boards as shown in Fig. 13.13. It is exposed from below to standard fire conditions according the Hydrocarbon curve, see Eq. 12.3. Heat transfer conditions are assumed according to Eurocode 1, i.e. $\varepsilon = 0.8$ and $h = 50$ W/(m^2 K). The thermal properties of steel and concrete are as given in Eurocode 2 and 3, respectively. The gypsum boards are 30 mm and have thermal properties according to Table 7.2.

A finite element discretization model was generated as shown in Figure 13.14a. Heat transfer inside the void between the steel web and the protection by radiation and convection was considered in the analysis.

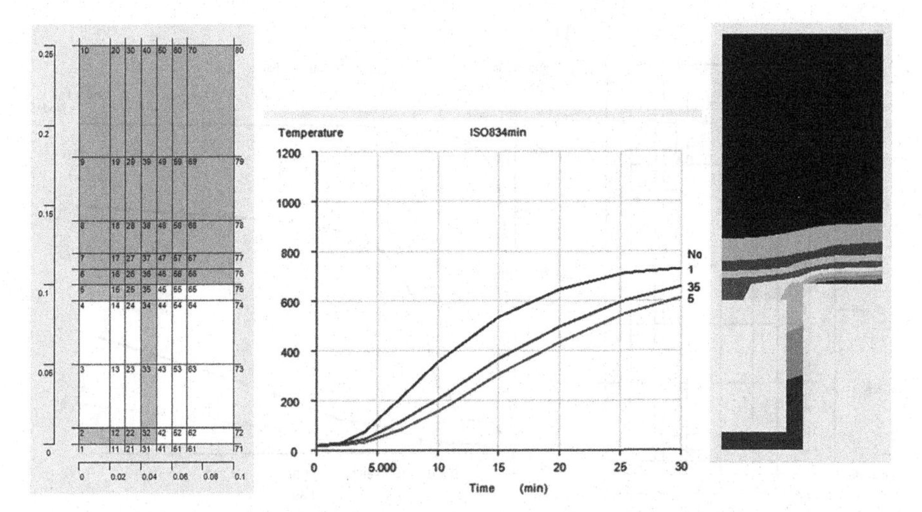

Fig. 13.12 Unprotected hollow section analysed by the finite element method. B/W plots from TASEF. (**a**) Finite element mesh of a symmetric half (**b**) Steel node temperatures vs. time (**c**) Temperature contours after 15 min

Fig. 13.13 An encased I-section steel (HE 300B) beam carrying a concrete slab. Slab thickness 160 mm, protection thickness 30 mm, steel height and width 300 mm, flange thickness 19 mm and web thickness 11 mm

The calculated temperature histories in the steel flanges are shown in Figure 13.14b. The vaporization of the water in the gypsum consumes a lot of heat as indicated by the enthalpy curve shown in Fig. 7.8. Therefore the uneven development of the temperature of the gypsum (curve #2). Notice also that the temperature difference between the minimum and maximum steel temperatures are in the order of 200 °C due to the cooling of the top flange by the concrete slab.

13.5.3 Unprotected I-Section Connected to a Concrete Structure

A bare HE300B steel section attached to a concrete structure as shown in Fig. 13.10 is exposed to standard fire conditions according ISO 834. Accurately calculated temperatures with the finite element code TASEF are shown in Fig. 13.15. Notice

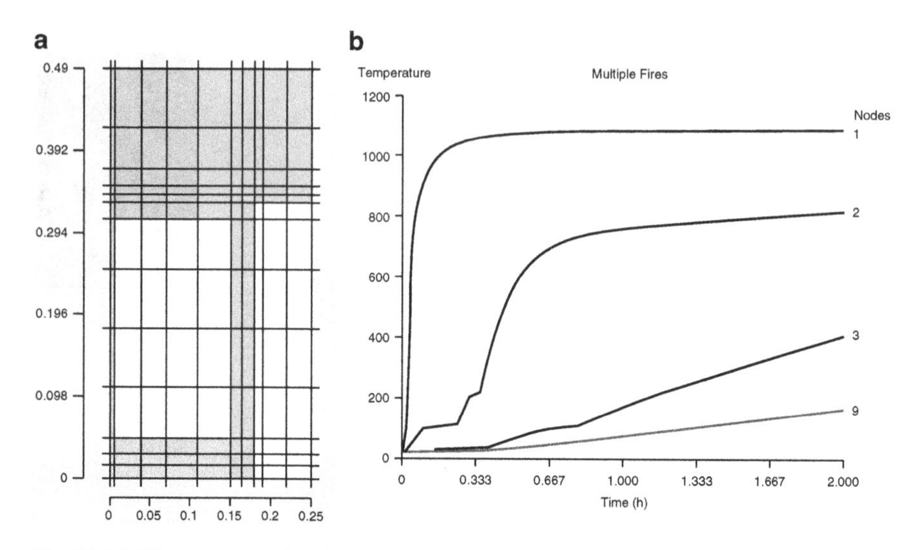

Fig. 13.14 I-beam protected with gypsum boards analysed by the finite element method. B/W plots from TASEF. (**a**) Finite element mesh of a symmetric half (**b**) Temperatures from above of gypsum surface, middle of gypsum, steel bottom and upper flanges

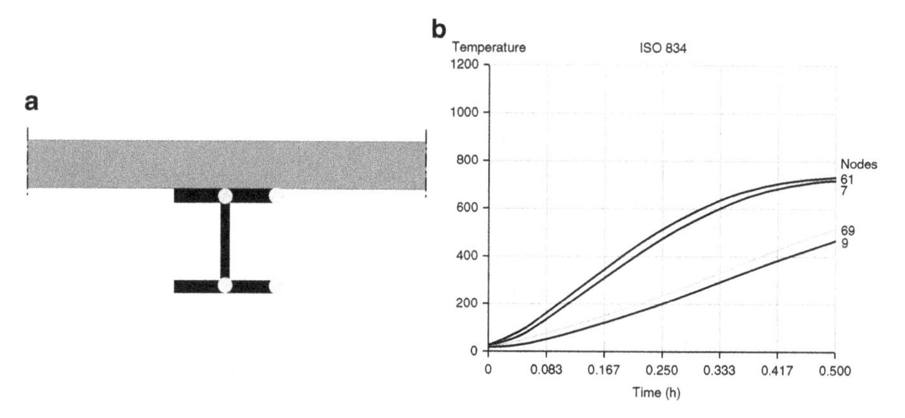

Fig. 13.15 Steel temperature development of the bottom and top flanges of assembly exposed to a standard ISO 834 time–temperature curve. Shadow effects and effects of cooling of the steel to the concrete are considered. (**a**) Points where calculated steel temperatures are shown in b) (**b**) Finite element calculated temperatures of flanges. B/W plot from TASEF

that when assuming lumped heat or uniform temperature a temperature of 500 °C is calculated after 15 min. In the finite element analysis this temperature is only reached by the bottom flange while the top flange attached to the concrete only reaches a temperature of 200 °C.

Chapter 14
Temperatures of Concrete Structures

Reinforced concrete structures are sensitive to fire exposure of mainly two reasons. They may be subject to explosive *spalling*, and they may lose their load-bearing capacity due to high temperatures. Spalling is particularly hazardous as it may occur more or less abruptly and unanticipated. It usually starts within 30 min of severe fire exposure. It may depend on several mechanisms or combinations thereof such as pore pressure, stresses due to temperature gradients, differences of thermal dilatation and chemical degradations at elevated temperatures. Reinforcement bars of steel lose their strength at temperature levels above 400 °C. Prestressed steel may even loose strength below that level. Concrete loose as well both strength and stiffness at elevated temperature.

As the spalling phenomenon is very complex and cannot be predicted with simple mathematical temperature models, it will not be further discussed here. For more detailed information regarding the fire spalling phenomenon see [47]. The procedures presented below presume that no spalling occur that could significantly influence the temperature development.

In Eurocode 2 [6] temperatures in fire-exposed structures may be obtained from tabulated values or by more or less advanced calculations. In the sections below thermal material properties as given in Eurocode 2 are reproduced and thereafter some simple approximate calculation methods are given in the following sections. For more general situations finite element calculations are needed.

14.1 Thermal Properties of Concrete

The conductivity of concrete decreases with rising temperature. It depends on concrete quality and type of ballast. For design purposes curves as shown in Fig. 14.1 may be used according to Eurocode 2 [6]. For more accurate calculations with alternative concrete qualities more precise material data may be obtained by measuring the thermal properties of the product in question, see Sect. 1.3.1.

U. Wickström, *Temperature Calculation in Fire Safety Engineering*,
DOI 10.1007/978-3-319-30172-3_14

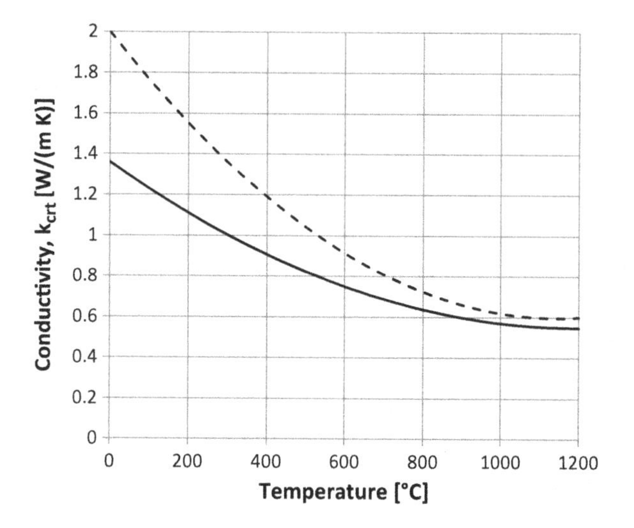

Fig. 14.1 Upper and lower limit of heat conductivity vs. temperature of normal weight concrete according to Eurocode 2 [6]

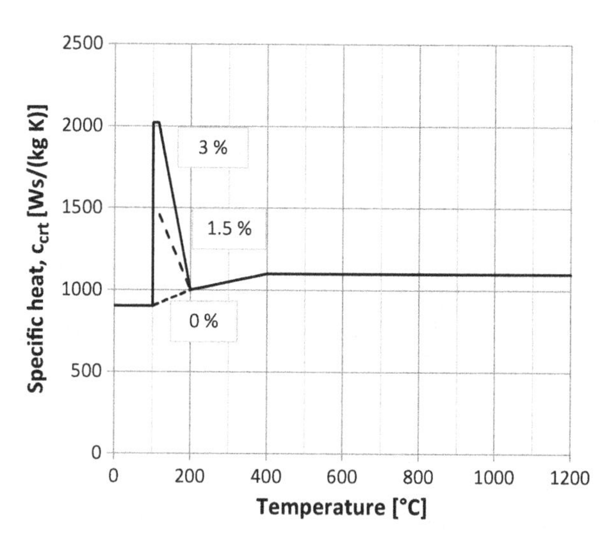

Fig. 14.2 Specific heat capacity of concrete vs. temperature at 3 different moisture contents, 0, 1.5 and 3 % for siliceous concrete according to Eurocode 2 [6]

The specific heat of dry concrete does not vary much with temperature. However, concrete structures always contain water which evaporates at temperatures above 100 °C constituting a heat sink (latent heat) as the vaporization process consumes a lot of heat. Thus the specific heat capacity for normal weight concrete according to Eurocode 2 has a peak at temperatures 100 and 200 °C as shown in Fig. 14.2.

The peak due to the latent heat involves a numerical challenge when calculating temperatures. Especially if the temperature range at which the vaporization of the moisture occurs becomes narrow, the peak becomes increasingly high. Then it can be advantageous to introduce the specific volumetric enthalpy as an input parameter as defined in Sect. 7.3.4. This formulation in combination with a forward difference

Table 14.1 Thermal properties of normal weight concrete according to Eurocode 2 [6] including the range of the conductivity between the upper and lower limits and the calculated volumetric enthalpy

			Moist. cont. 0 %		Moist. cont. 1.5 %		Moist. cont. 3 %	
T	k	ρ	c	e	c	e	c	e
[°C]	[W/(m K)]	[kg/m^3]	[(Ws)/(kg K)]	[(Wh)/m^3]	[(Ws)/(kg K)]	[(Wh)/m^3]	[(Ws)/(kg K)]	[(Wh)/m^3]
0	1.36–2.00	2300	900	0	900	0	900	0
20	1.33–1.95	2300	900	11,500	900	11,500	900	11,500
100	1.23–1.77	2300	900	57,500	900	57,500	900	57,500
115	1.21–1.73	2300	915	66,197	1470	71,587	2020	76,858
200	1.11–1.55	2254	1000	117,154	1000	137,313	1000	157,220
400	0.91–1.19	2185	1100	244,613	1100	264,772	1100	284,678
1200	0.55–0.60	2024	1100	739,368	1100	759,527	1100	779,434

See also diagram in Fig. 14.3

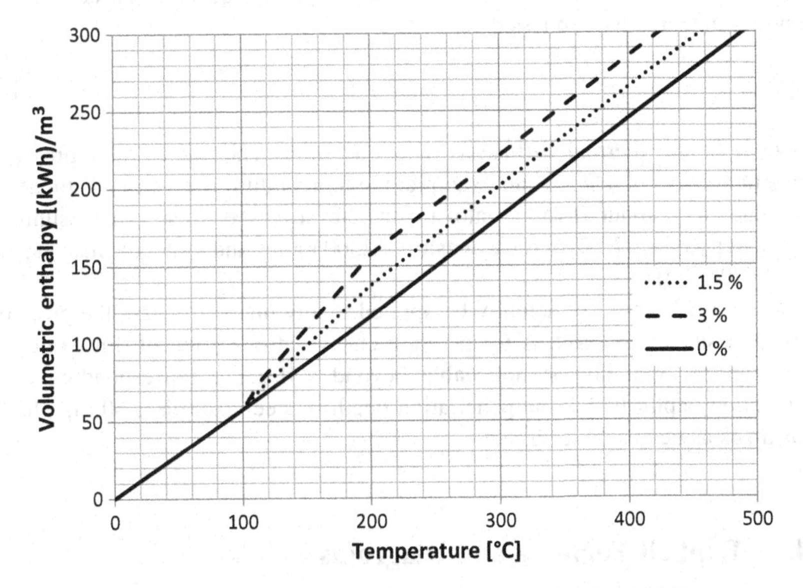

Fig. 14.3 Volumetric enthalpy of concrete for moisture contents 0, 1.5 and 3.0 % vs. temperature based on density and specific heat capacity according to Eurocode 2 [6] (see Table 14.1)

time integration scheme is used in the computer code TASEF [14]. Table 14.1 and Fig. 14.2 show calculated values of the specific volumetric enthalpy vs. temperature starting at 0 °C based on specific heat and density values given in Eurocode 2 [6] for normal concrete. Notice that no consideration is given to the latent heat of the water before it vaporizes under 100 °C. This is generally an acceptable approximation for normal weight concrete but not for many other materials which may contain much higher percentages of moisture.

The Eurocode on concrete (EN 1992-1-2) states that the emissivity related to concrete surfaces should be taken as 0.7. The Eurocode 1 on actions (EN 1991-1-2) gives the convective heat transfer coefficient when simulating fully developed fires to be assumed equal to 25 W/(m^2 K). (In general the assumed values of the surface emissivity and convective heat transfer coefficient have only marginal influence on calculated temperatures inside concrete structures.)

14.2 Penetration Depth in Semi-infinite Structures

Concrete is a material with relatively high density and low conductivity. It therefore takes a long time for heat to penetrate into the structure and raise its temperature, or in other words it takes time before a temperature change at one point is noticeable at another point. Thus in many cases a concrete structure may be assumed semi-infinite. In Sect. 3.2.1.1 it is shown that temperature change at the surface will only be noticeable at a depth δ less than

$$\delta = 2.8\sqrt{\alpha \cdot t} \tag{14.1}$$

where α is the thermal diffusivity and t is time. The value 2.8 represents a temperature rise of 1 %. As an example, the temperature rise can be estimated to penetrate only about 0.15 m into a concrete structure after 1 h (assuming a conductivity of a 1.7 W/(m K), a density of 2300 kg/m^3 and a specific heat capacity of 900 J/(kg K)).

Penetration depth can actually be applied to any material where the properties may be assumed constant. A temperature change at one point of, for example, a steel member will not be noticeable beyond a distance corresponding to the penetration depth. In 1 h the penetration depth in steel exceeds 0.60 m, which is four times as deep as in concrete.

14.3 Explicit Formula and Diagrams

In general numerical procedures such as finite element methods are needed to calculate temperature in concrete structures. A 1-D configuration of a concrete is shown in Fig. 14.4.

Wickström [48, 49] has, however, shown that in 1-D cases may the temperature inside concrete structures exposed to standard fire conditions according to ISO 834 and heat transfer condition according to Eurocode 1 (Eq. 12.11) be obtained from explicit formula and diagrams. The diagrams as shown in Figs. 14.5 and 14.6 were then obtained by comparisons with numerous finite element calculations. They yield concrete temperatures which coincide with the temperatures obtained with the accurate numerical methods within a few per cent in the interesting area of

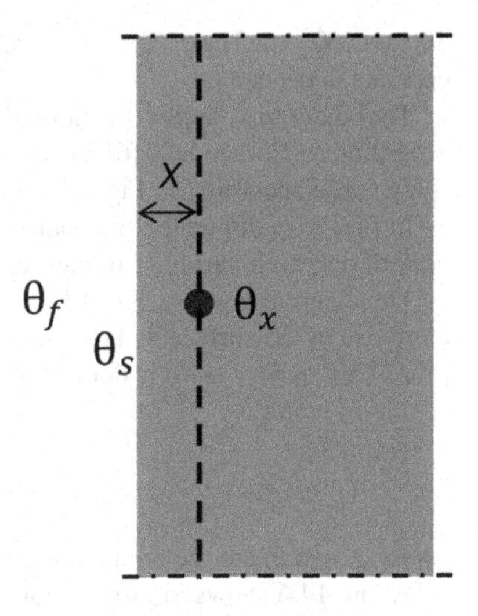

Fig. 14.4 Definitions of temperature rises of a 1-D thick concrete wall exposed to fire on one side

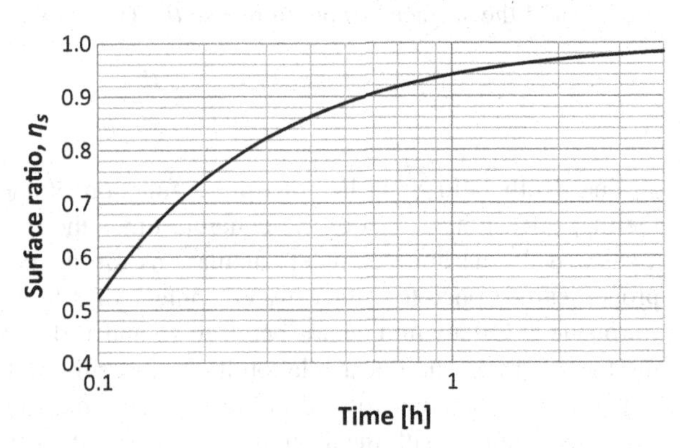

Fig. 14.5 The surface ratio η_s vs. time for a normal weight concrete with thermal properties according to Eurocode 2 [6] exposed to standard fire conditions according to ISO 834

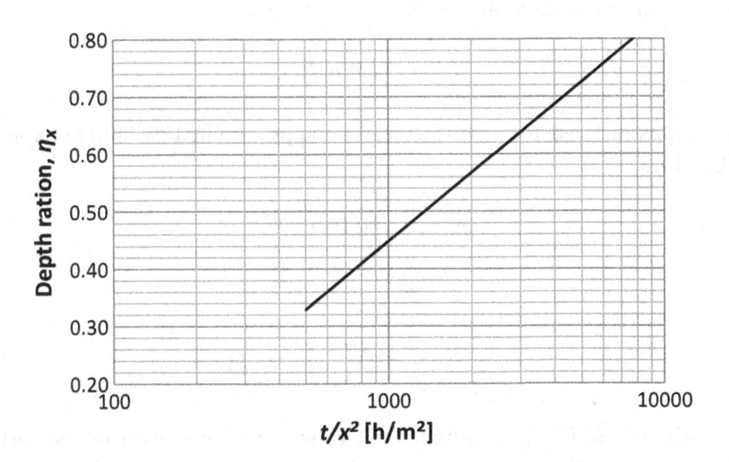

Fig. 14.6 The *in depth ratio* η_x vs. time divided by depth squared t/x^2 *for* normal weight concrete with thermal properties according to Eurocode 2 [7] exposed to standard ISO 834 fire conditions. Calculations are made assuming lower limit of the conductivity as shown in Fig. 14.1

300 to 600 °C. These diagrams are therefore very handy to use when a quick estimate is needed.

The diagrams apply to normal weight concrete with thermal properties according to Eurocode 2 [6] as shown in Table 14.1 assuming the lower conductivity curve according to Fig. 14.1 and a moisture content of 1.5 %.

In [48] it is shown that the same type of diagrams can be used more generally considering both various parametric fires and various material properties.

The diagram given in Fig. 14.5 shows the ratio η_s between the concrete temperature rise of the surface and the standard fire temperature according to ISO 834 vs. time. This *surface ratio* is defined as

$$\eta_s = \frac{\theta_s}{\theta_f} \tag{14.2}$$

where θ_s and θ_f are the temperature rise of the surface and the fire, respectively.

Figure 14.6 shows in turn the ratio between the internal temperature rise θ_x at a depth x and the surface temperature rise θ_s. This *depth ratio* is defined as

$$\eta_x = \frac{\theta_x}{\theta_s} \tag{14.3}$$

The depth ratio η_x is in principle a function of the Fourier number, i.e. the thermal diffusivity $k/(c\rho)$ of the concrete times the fire duration t over the depth x squared. In the finite element calculations for developing the diagrams thermal properties of concrete with a water content of 1.5 % are assumed according to Eurocode 2. Calculation depths between 25 and 100 mm were used when developing the diagram. The linear relation in the logarithmic-linear diagram as shown in Fig. 14.6 was then constructed. It yields approximate temperatures slightly higher than was obtained with the accurate finite element calculations.

The internal concrete temperature may now be written as

$$T_x = \eta_s \eta_x T_f \tag{14.4}$$

The graphs in Figs. 14.5 and 14.6 can be approximated by simple expressions. Thus Eq. 14.5

$$\eta_s = 1 - 0.060\, t^{-0.90} \tag{14.5}$$

and

$$\eta_x = 0.172 \ln\left(\frac{t}{x^2}\right) - 0.74 \tag{14.6}$$

respectively, where t is time in hours and x distance in metres from the surface.

Then in summary for standard fire exposure according to ISO 834 and normal weight concrete according to Eurocode 2 [7] (see Sect. 14.1) a very simple closed form solution may be obtained. Thus the surface temperature rise is

$$\theta_s = \left[1 - 0.060\, t^{-0.90}\right] \cdot [345 \cdot \log(480t + 1)]\ [^\circ\mathrm{C}] \tag{14.7}$$

The internal temperatures at arbitrary times and depths are obtained by inserting Eqs. 14.5 and 14.6 into Eq. 14.4 of a structure initially at 20 °C then becomes:

$$T_{x,t} = \left[1 - 0.060\, t^{-0.90}\right] \cdot \left[0.172 \ln\left(\frac{t}{x^2}\right) - 0.74\right] \cdot [345 \cdot \log(480t + 1)] + 20\ [^\circ\mathrm{C}] \tag{14.8}$$

A diagram based on Eqs. 14.7 and 14.8 is shown in Fig. 14.7 including the standard ISO 834 fire curve. The graphs are limited between 200 and 700 °C. Outside that range Eq. 14.8 is not valid.

As an illustration the temperature in a slab of normal-weight concrete is calculated at a depth of 4 cm when exposed to an ISO 834 standard fire for 1 h. At first η_s is obtained from Fig. 14.5 to be 0.97 at $t = 1$ h. Then for $t/x^2 = 2.0/(0.04)^2 = 1250$ h/m^2 and Eq. 14.5 or Fig. 14.6 yields approximatively $\eta_x = 0.49$. As the standard fire temperature rise after 1 h is 1029 °C, the concrete surface temperature rise is obtained from Eq. 14.8 as $0.97 \cdot 1029 = 998$ °C and Eq. 14.8 yields the temperature rise at a depth of 4 cm to be $T_x = 0.97 \cdot 0.49 \cdot 1029 + 20\ ^\circ\mathrm{C} = 509\ ^\circ\mathrm{C}$. Alternatively a direct reading of Fig. 14.7 yields a $T_x = 500$ °C which coincides very well with an accurate finite element calculation.

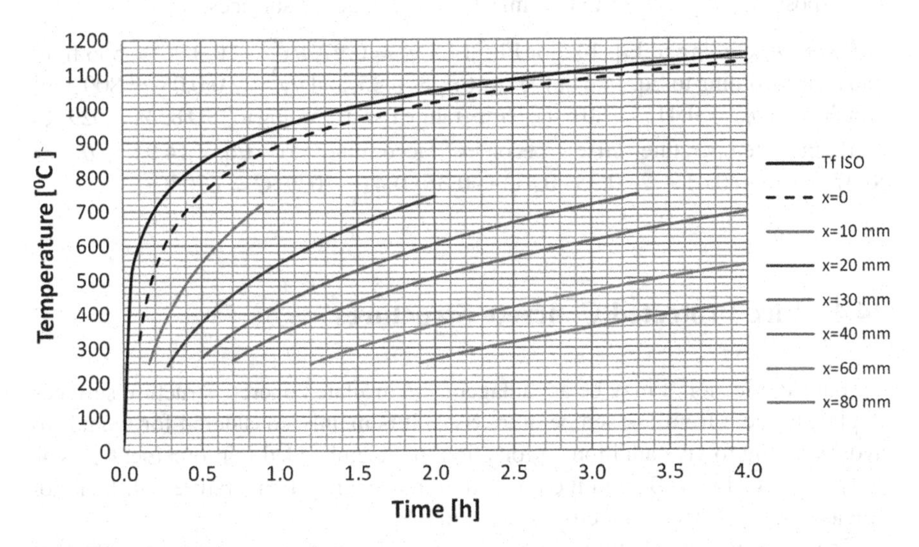

Fig. 14.7 Temperature in concrete based on Eq. 14.8 in the range of 200 to 700 °C at various depths when exposed to the standard ISO 834 fire curve. The temperatures of the exposure curve and the surface are given as well

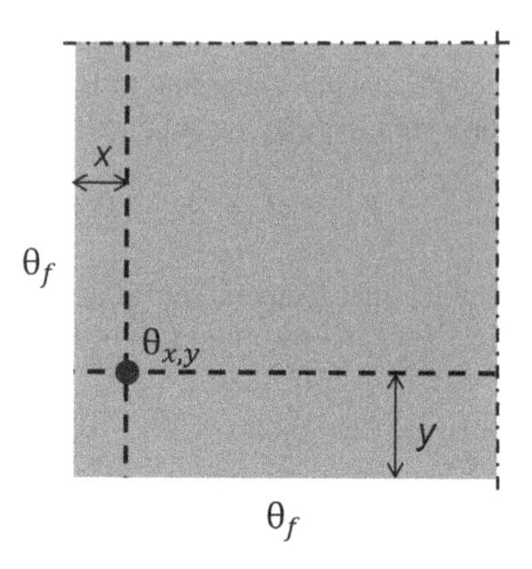

Fig. 14.8 Definitions of temperature rises at a 2-D concrete corner exposed to fire from two sides

Also the temperature rise near 2-D corners exposed to ISO 834 standard fires may be calculated using the approximations above [48]. Thus the temperature at a point at distances x and y, respectively, from the exposed surfaces (see Fig. 14.8) may be calculated as

$$\theta_{x,y} = \left[\eta_s \cdot \left(\eta_x + \eta_y - 2 \cdot \eta_x \eta_y\right) + \eta_x \eta_y\right] \cdot \theta_f \tag{14.9}$$

where η_s is the surface ratio according to Eq. 14.5 or Fig. 14.5, and η_x and η_y are the depth ratios in the x and y directions, respectively, according to Eq. 14.6 or Fig. 14.6.

Example 14.1 Calculate the temperature in a rectangular concrete beam after 2.0 h fire exposure at a point 60 and 50 mm from the exposed surfaces.

Solution According to Eq. 14.5 or Fig. 14.5 $\eta_s = 0.97$, $t/x^2 = 2.0/0.06^2 = 556$ h/m^2 and then according to Eq. 14.6 or Fig. 14.6 $\eta_x = 0.35$, and $t/y^2 = 2.0/0.05^2 = 800$ h/m^2 which yields $\eta_y = 0.41$. At 2.0 h the temperature rise according to ISO 834 is 1029 °C, and the temperature rise becomes according to Eq. 14.9 $\theta_{x,y} = [0.97 \cdot (0.35 + 0.41 - 2 \cdot 0.35 \cdot 0.41) + 0.35 \cdot 0.41] \cdot 1029\,°\mathrm{C} = 620\,°\mathrm{C}$.

14.4 Fire Protected Concrete Structures

In some application it may be advantageous to insulate concrete structure surfaces to prevent them from fast temperature rises. It is mainly considered for tunnels to avoid spalling to give additional protection to the embedded reinforcement bars as shown in Fig. 14.9a. Behind the protection the concrete temperature will then not rise as quickly as when directly exposed to fire.

There are in principle three types of passive fire protections used for protection of tunnels, namely spraying with cementitious mortar, lining with non-combustible

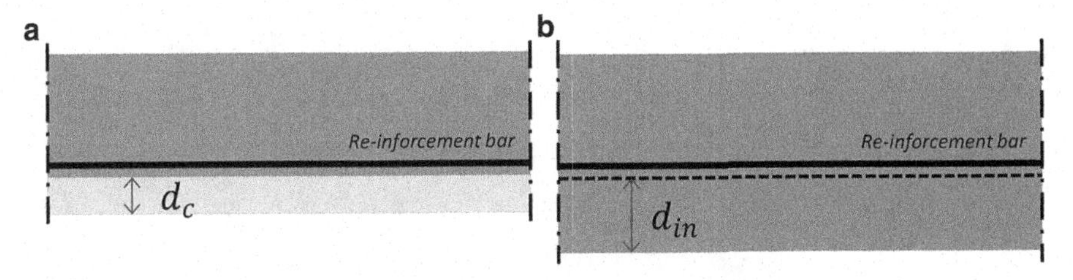

Fig. 14.9 The protection of a concrete structure layer with a thickness d_{in} gives an equivalent thermal protection as a concrete layer with a thickness $d_c = k_c d_i / k_i$. (**a**) Concrete slab fire protected from below (**b**) Concrete layer providing equivalent thermal protection

boards and lining with concrete containing polypropylene fibres. For more information on fire dynamics in tunnels see [50] and on concrete in tunnels see [51]. A simple way of estimating how much thermal protection an insulation provides in terms of concrete thickness based on finite element calculations [19] has been suggested by Wickström and Hadziselimovic [52].

They showed that the same effect is approximately obtained when the thermal resistance of the insulation is the same as that for the concrete layer. Thus the equivalent concrete layer thickness can be calculated as

$$d_c = \frac{k_c \cdot d_{in}}{k_{in}} \tag{14.10}$$

where d is thickness and k conductivity, and the indices in and c stand for insulation and concrete, respectively.

As an example a 10 mm board of vermiculite with a thermal conductivity of 0.2 W/(m K) corresponds to a concrete protection layer of 50 mm assuming the concrete has a conductivity of 1.0 W/m K for the temperature interval considered. This could mean considerable savings in both weight and space for a concrete structure.

Chapter 15
Temperature of Timber Structures

Modelling the thermal behaviour of wood is complicated as phenomenas such as moisture vaporization and migration, and the formation of char have decisive influences on the temperature development within timber structures. Nevertheless it has been shown that general finite element codes can be used to predict temperature in, for example, fire-exposed cross sections of glued laminated beams [53], provided, of course, that apparent thermal material properties and appropriate boundary conditions are used. Other specialized numerical codes for timber structures have been developed, e.g. by Fung [54] and Gammon [55]. A comprehensive collection of papers on timber in fire is listed in [56].

15.1 Thermal Properties of Wood

Both density and moisture content affect the thermal properties of wood. In the literature a wide range of values are given. In the SFPE Handbook of Fire Protection Engineering [4], the following equation is given for the conductivity in W/(m K) as

$$k = \rho \cdot (194.1 + 4.064 \cdot u) \cdot 10^{-6} + 18.64 \cdot 10^{-3} \tag{15.1}$$

and for the specific heat capacity of dry wood in Ws/(kg K) as

$$c = 103.1 + 3.867\,\overline{T} \tag{15.2}$$

where ρ is the density based on volume at current moisture content and oven-dry weight (kg/m^3), u the moisture content (per cent by weight) and $\overline{T}$ is the temperature (K). These values are mainly developed for temperatures below 100 °C. For higher temperatures the latent heat for the vaporization of free water must be considered as

U. Wickström, *Temperature Calculation in Fire Safety Engineering*,
DOI 10.1007/978-3-319-30172-3_15

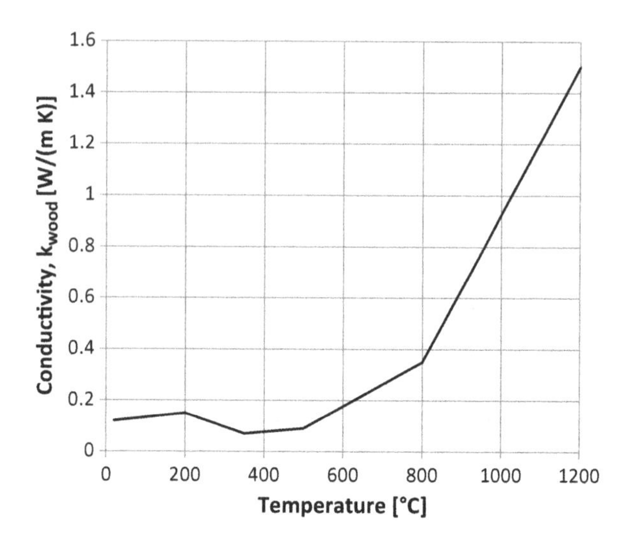

Fig. 15.1 Temperature vs. conductivity for wood and the char layer according to Eurocode 5. See also Table 15.1

Table 15.1 Temperature vs. conductivity for wood and the char layer according to Eurocode 5

Temperature [°C]	20	200	350	500	800	1200
Conductivity [W/(m K)]	0.12	0.15	0.07	0.09	0.35	1.5

See also Fig. 15.1

for concrete, see Sect. 14.1. At equilibrium in "normal" conditions (20 °C and a relative air moisture content of 65 %) wood contains about 12 % by weight of water.

According to Eurocode 5 (EN 1965-1-5) Annex B the conductivity is recommended to be as shown in Fig. 15.1 and Table 15.1, and the specific heat capacity as shown in Fig. 15.2. These material properties are limited structures exposed to standard fire exposure according to ISO 834 or EN 1363-1, see [56].

Figure 15.3 shows the specific volumetric enthalpy based on Table 15.2 for wood with a density of 450 kg/m^3 and a moisture content of 12 %.

The thermal properties of wood are in general very uncertain and it is very hard to find reliable data in the literature. For approximate calculations it is here recommended to use a constant conductivity of 0.13 W/(m K) independent of moisture content and a specific heat of 2000 W/(kg K) for dry wood with additions for the sensitive and latent heat of water as described for concrete in Sect. 14.1.

15.2 Charring Depth According to Eurocode 5

Simple estimations of load-bearing capacities of timber members are according to Eurocode 5 [7] made in two steps. First a residual cross section is calculated by removing the char layers entering from fire-exposed surfaces. Then the mechanical properties of the residual cross section are calculated based on the remaining virgin wood. This procedure is called Reduced Cross-Section Method or Effective Cross-Section Method.

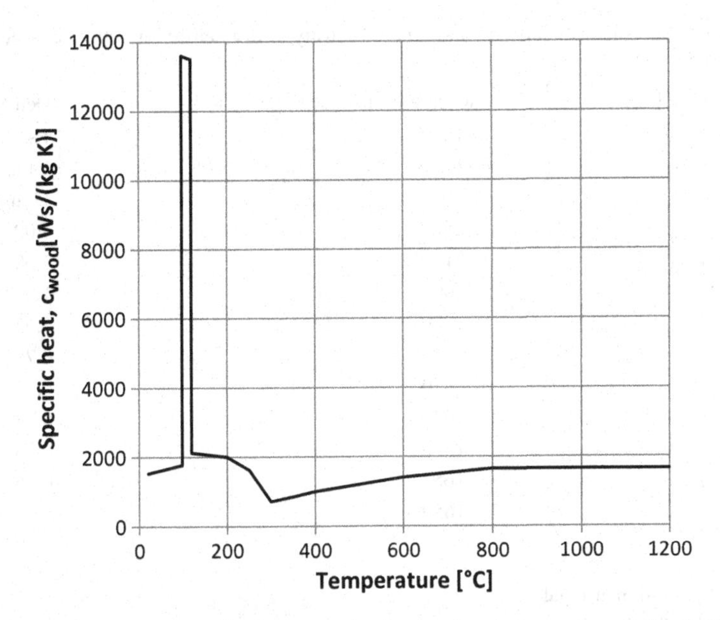

Fig. 15.2 Temperature vs. specific heat capacity for wood and charcoal according to Eurocode 5. The peak at 100 °C corresponds to the heat of vaporization of 12 % by weight of water

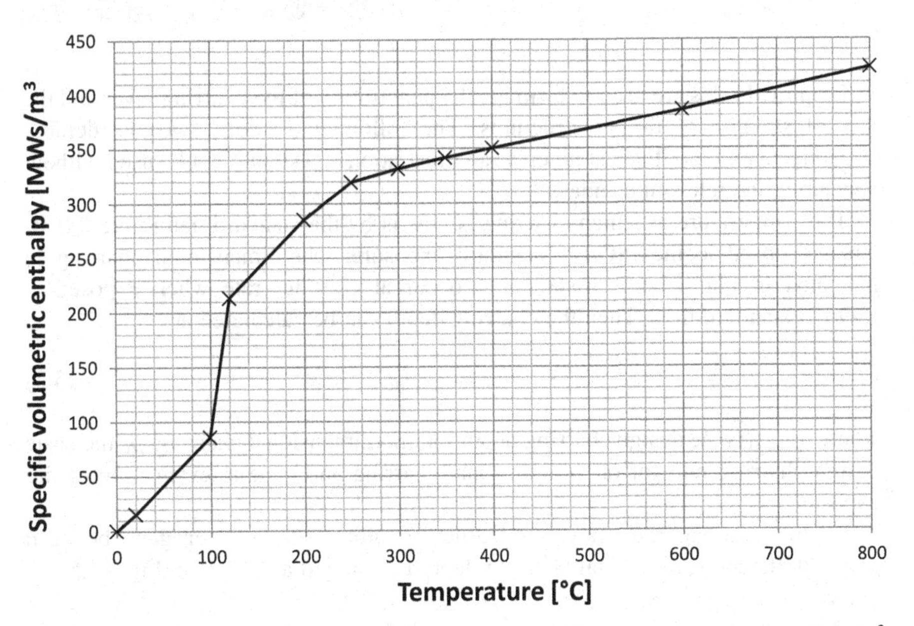

Fig. 15.3 Example of specific volumetric enthalpy in MWs/m^3 for wood with a density 450 kg/m^3 and a moisture content by weight of 12 %

Table 15.2 Specific heat capacity and ratio of density to dry density of softwood according to Eurocode 5

Temperature [°C]	Specific heat capacity [J/(kg K)]	Density ratio
20	1530	1 + u/100
99	1770	1 + u/100
99	13,600	1 + u/100
120	13,500	1.00
120	2120	1.00
200	2000	1.00
250	1620	0.93
300	710	0.76
350	850	0.52
400	1000	0.38
600	1400	0.28
800	1650	0.26
1200	1650	0

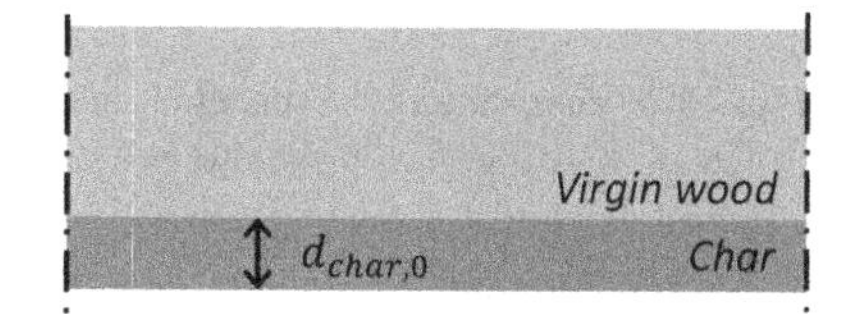

Fig. 15.4 One-dimensional charring (fire exposure on one side) [7]

Empirical rules are used to estimate the penetration of the charring layer and the loss of strength of timber structures. The following section is a considerably abbreviated extract. It is just given as an illustration and should not be used without consulting the relevant standard.

The temperature at which charring begins is by the standard definition 300 °C when exposed to the ISO/EN standard exposure. One-dimensional charring as indicated in Fig. 15.4 is assumed to occur at constant rate when exposed to ISO/EN standard fires. Then the charring depth can be calculated as

$$d_{char,0} = \beta_0 t \tag{15.3}$$

where $d_{char,0}$ is the design charring depth for one-dimensional charring, β_0 the basic design charring rate for one-dimensional charring and t the relevant time of fire exposure.

When including the effects of corner roundings, fissures or gaps between adjacent elements, a notional charring depth is assumed as shown in Fig. 15.5:

$$d_{char,n} = \beta_n t \tag{15.4}$$

where $d_{char,n}$ is the notional design charring depth.

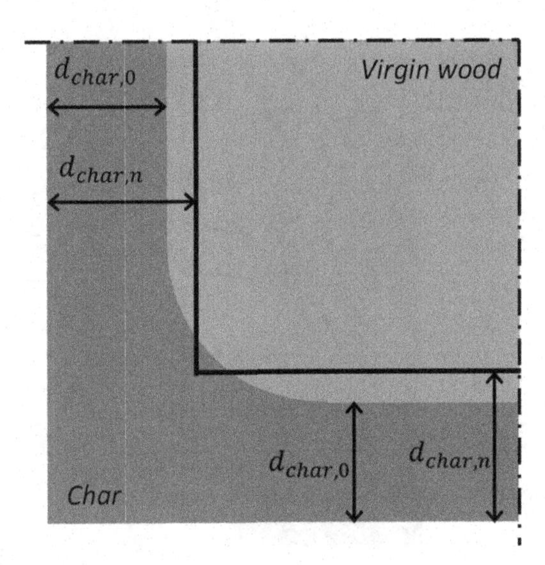

Fig. 15.5 Charring depth $d_{char,0}$ for one-dimensional charring and notional charring depth $d_{char,n}$ [7]

Table 15.3 Design charring rates β_0 and β_n of timber, LVL, wood panelling and wood-based panels [7]

	β_0 [mm/min]	β_n [mm/min]
(a) Softwood and beech		
Glued laminated timber with a characteristic density of ≥ 290 kg/m^3	0.65	0.7
Solid timber with a characteristic density of ≥ 290 kg/m^3	0.65	0.8
(b) Hardwood		
Solid or glued laminated hardwood with a characteristic density of ≥ 290 kg/m^3	0.65	0.7
Solid or glued laminated hardwood with a characteristic density of ≥ 450 kg/m^3	0.50	0.55
(c) LVL (Laminated Veneer Lumber)		
With a characteristic density of ≥ 480 kg/m^3	0.65	0.7
(d) Panels		
Wood panelling	0.9[a]	–
Plywood	1.0[a]	–
Wood-based panels other than plywood	0.9[a]	–

[a]The values apply to a characteristic density of 450 kg/m^3 and a panel thickness of 20 mm or more

For initially unprotected surfaces of timber design charring rates β_0 and β_n are given in Table 15.3.

More details on how to estimate charring depths are given in Eurocode 5 [7].

Timber members may be protected by fire claddings or other protection materials to delay the start of charring. Rules on how to calculate the start of charring of protected timber are given in Eurocode 5 [7].

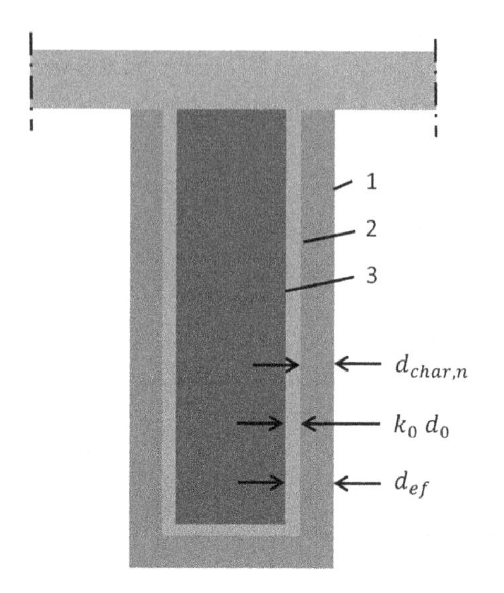

Fig. 15.6 Definition of residual cross section and effective cross section [7]

Table 15.4 Determination of k_0 in Eq. 15.5 for unprotected surfaces with t in minutes. From Eurocode 5

Time	k_0
$t < 20$ min	$t/20$
$t \geq 20$ min	1.0

When determining the cross-sectional mechanical properties, an effective cross section should be calculated by reducing the initial cross section by the effective charring depth d_{ef} (see Fig. 15.6). Then

$$d_{ef} = d_{char,n} + k_0 d_0 \tag{15.5}$$

where $d_0 = 7\,\text{mm}$ is the zero-strength layer. $d_{char,n}$ is determined according to Eq. 15.4. For unprotected surfaces, k_0 should be determined according to Table 15.4.

It is then only the effective part that shall be accounted for when calculating the mechanical properties of a cross section. When using the Reduced Cross-Section Method it is assumed that the effective cross section has ambient material properties. All losses in strength and stiffness are compensated by the zero-strength layer.

Example 15.1 A glued laminated beam (300 mm by 500 mm high) of pine (softwood) is exposed on three sides to a standard EN/ISO curve. Calculate the effective cross section after a fire exposure of 60 min.

Solution The effective charring depth d_{ef} can be calculated according to Eq. 15.5 and $d_{char,n}$ from Eq. 15.4. Thus $\mathrm{d_{ef}} = 0.7 \cdot 60 + 1.0 \cdot 7\,\mathrm{mm} = 49\,\mathrm{mm}$ and the remaining effective cross section becomes (500-49) mm by (300-2 · 49) mm equal to 451 mm by 202 mm.

End-user-friendly information for designers including examples can be found in [58].

Review Questions

Chapter 1

1. What is the difference between heat and temperature?
2. Which are the three modes of heat transfer?
3. What is the "driving force" of heat transfer?
4. Write Fourier's law of heat conduction.
5. Which are the three types of boundary condition, 1, 2 and 3?
6. Which of the three types of boundary conditions is the most common in FSE?
7. What is an adiabatic surface?
8. How is heat transferred from the gas phase to a solid surface?
9. Write the equation for a convection boundary condition.
10. What is absorbed radiation?
11. Write the expression for the emitted radiation from a surface according to the Stefan–Boltzmann law.
12. What is net radiation heat flux?
13. What is incident black body radiation temperature or just the black body temperature T_r?
14. What is a mixed boundary condition?
15. How is the fire boundary condition normally written in standards on fire resistance of structures?
16. Write the heat conduction equation in 1-D.
17. Explain the parameters in the heat conduction equation.
18. What is thermal diffusivity?
19. What is specific volumetric enthalpy?
20. What is thermal inertia and why does it vary so much for various materials?
21. What happens to steel properties at elevated temperatures?
22. Which is the main problem with concrete structures exposed to severe fire conditions?
23. Why can wooden structures resist fires relatively well?

U. Wickström, *Temperature Calculation in Fire Safety Engineering*,
DOI 10.1007/978-3-319-30172-3

Chapter 2

24. Draw the temperature distribution of a wall under steady-state conditions. Assume constant thermal parameters, 3rd kind of boundary condition on one side and 1st kind on the other side.
25. What is the total thermal resistance of a wall with a thickness L, a conductivity k and heat transfer coefficients h at the bounding surfaces?

Chapter 3

26. What is the meaning of lumped-heat-capacity or uniform temperature?
27. Write and explain the heat balance equation where lumped-heat-capacity is assumed.
28. Under what conditions can an analytical solution be derived for the uniform temperature of a body exposed to elevated gas temperature?
29. Give two examples when uniform temperatures can be assumed in fire safety engineering.
30. What is meant by a semi-infinite body?
31. Show in the diagram how the temperature profile develops in a semi-infinite body experiencing a sudden temperature rise at the surface.
32. What is penetration depth?
33. Which is the material parameter group governing the temperature development of semi-infinitely thick bodies with a prescribed surface temperature?
34. Which is the material parameter group governing the surface temperature development of semi-infinitely thick bodies with a prescribed heat flux at the surface?
35. Which is the parameter group governing the surface temperature development of semi-infinitely thick bodies exposed to a prescribed gas temperature and heat transfer coefficient?

Chapter 4

36. Why is radiation so important in fire safety engineering?
37. What is the radiation heat transfer coefficient and how can it be calculated?
38. What is adiabatic surface temperature and how is it defined?
39. Which parameters are needed to calculate the adiabatic surface temperature?

Chapter 5

40. What is the resultant emissivity between two parallel plates?
41. What is a view factor?
42. What is an absorption or emission coefficient?
43. Which parameters determine the emissivity of a flame?

Chapter 6

44. What governs heat transfer by convection?
45. Which are the two principal ways of inducing air flow?
46. Which are the two principal types of flow patterns?
47. Which air properties govern the magnitude of the convection heat transfer coefficient?
48. What is effective or apparent thermal conductivity in enclosed species?

Chapter 7

49. Write the heat balance equation in numerical form of a body exposed to incident radiation and a gas temperature assuming lumped heat?
50. Write the transient heat balance equation in the matrix form. Describe the components.
51. How can the equation be solved? What are the advantages and disadvantages of explicit and implicit methods?
52. What is specific volumetric enthalpy?
53. Specific volumetric enthalpy of a wet material has three components at temperatures above vaporization. Which?

Chapter 8

54. According to thermal ignition theory there are formulas to calculate time to reach critical temperatures for thin and thick solids, respectively. Which parameters are governing in the two cases?
55. What is the critical incident radiation heat flux?

Chapter 9

56. Describe the function of a thermocouple.
57. When measuring gas temperature with thermocouples there are two main error sources, which?
58. How should a thermocouple be designed to measure gas temperatures accurately?
59. What does a heat flux meter of Gardon gauge or Schmidt-Boelter gauge measure?
60. Why is the PT larger than a thermocouple?
61. A plate thermometer measures approximatively the adiabatic surface temperature of a relatively large surface. Why not exactly?
62. How can incident thermal radiation be calculated based on plate thermometer measurements?
63. How can adiabatic surface temperatures be calculated based on plate thermometer measurements?

Chapter 10

64. Which are the four main components of the heat balance equation of a fully developed compartment fire?
65. Which is the driving force of the gas flow in a one-zone model?
66. What is the difference between ultimate fire temperature and maximum temperature?

Chapter 11

67. What is the difference between a one-zone and a two-zone model? When are they applicable?
68. Which is the driving force of the gas flow in a two-zone model?

Chapter 12

69. Which three major steps consist of a fire design or analysis process of?
70. What is the meaning of the gamma factor?
71. How is the gamma factor influenced by the opening factor and the thermal inertia of the compartment boundaries?
72. What determines the duration of a parametric fire?

Chapter 13

73. What characterizes the thermal properties of steel?
74. What is “lumped-heat-capacity”?
75. What is “section factor” or “shape factor” of a steel section?
76. What is meant by heavily protected steel structures?
77. What is “shadow effect”?

Chapter 14

78. What characterizes the thermal properties of concrete in comparison with steel and insulation materials?
79. What is the surface temperature of a thick concrete wall after 1 h fire exposure according to the ISO 834 standard curve. Use the diagram in Fig. 14.5.
80. What is the temperature 3 cm into a thick concrete structure after one hour fire exposure according to the ISO 834 standard curve? Use the diagrams in Figs. 14.5 and 14.6
81. Calculate the same temperatures as in the two questions above but use Eq. 14.7 and Eq. 14.8, respectively?
82. How can the insulation of a concrete structure be considered in terms of equivalent concrete thickness?

Chapter 15

83. What needs to be considered particularly when calculating temperature in timber structures?
84. How is the thermal conductivity of wood in comparison to steel and concrete?
85. How are simple estimations made of load-bearing capacities of timber members according to for example Eurocode 5?

References

1. Holman J (1986) Heat transfer, 6th edn. McGraw-Hill Book Company, New York
2. Incropera FP, deWitt DP (1996) Fundamentals of heat and mass transfer, 4th edn. Wiley, New York
3. The European Committee for Standardisation, CEN (2005) EN 1993-1-2, Eurocode 3: design of steel structures—general rules—structural fire design. The European Committee for Standardisation, CEN, Brussels
4. DiNenno PJ et al (eds) (2008) SFPE handbook of fire protection engineering, 4th edn. National Fire Protection Association, Quincy
5. Hurley M et al (eds) (2015) SFPE handbook of fire protection engineering, 5th edn. SFPE, Gaithersburg
6. The European Committee for Standardisation, CEN (2004) EN 1992-1-2, Eurocode 2: design of concrete structures—general rules—structural fire design. The European Committee for Standardisation, CEN, Brussels
7. The European Committee for Standardisation, CEN (2004) EN 1995-1-2, Eurocode 5: design of timber structures—general rules—structural fire design. The European Committee for Standardisation CEN, Brussels
8. Flynn D (1999) Response of high performance concrete to fire conditions: review of thermal property data and measurement techniques, NIST GCR 99-767. National Institute of Standards and Technology, Gaithersburg
9. Adl-Zarrabi B, Boström L, Wickström U (2006) Using the TPS method for determining the thermal properties of concrete and wood at elevated temperature. Fire Mater 30:359–369
10. McGrattan KB, Hostikka S, Floyd JE, Baum HR, Rehm RG (2015) Fire dynamics simulator, technical reference guide, vol 1: mathematical model. NIST Special Publication 1018-1. National Institute of Standards and Technology, Gaithersburg
11. Siegel R, Howell J (1972) Thermal radiation heat transfer. McGraw-Hill Book Company, New York
12. Försth M, Roos A (2010) Absorptivity and its dependence on heat source temperature and degree of thermal breakdown. Fire Mater 35(5):285–301
13. Drysdale D (1992) An introduction to fire dynamics. Wiley-Interscience, New York
14. Wickström U (1979) TASEF-2—a computer program for temperature analysis of structures exposed to fire. Doctoral thesis, Lund Institute of Technology, Department of Structural Mechanics, Report NO. 79-2, Lund
15. Pitts DR, Sissom LE (1977) Schaums outline of theory and problems of heat transfer. Schaum's outline series, ISBN 0-07-050203-X, McGraw-Hill book company, New York

U. Wickström, *Temperature Calculation in Fire Safety Engineering*,
DOI 10.1007/978-3-319-30172-3

16. Jiji LM (2009) Heat conduction, 3rd edn. Springer, Berlin
17. Sjöström J, Wickström U (2014) Superposition with Non-linear Boundary Conditions in Fire Sciences, Fire Technology. Springer, New York
18. SFPE Standard on Calculation Methods to Predict the Thermal Performance of Structural and Fire Resistive Assemblies (draft)
19. Sterner E, Wickström U (1990) TASEF—temperature analysis of structures exposed to fire. SP Report 1990:05, SP Technical Research Institute of Sweden, Boras
20. Fransén JM, Kodur VKR, Mason J (2000) User's manual of SAFIR 2001. A computer program for analysis of structures submitted to fire. University of Liege, Belgium
21. SFPE (2015) SFPE engineering standard on calculation methods to predict the thermal performance of structural and fire resistive assemblies (SFPE S.02 2015). SFPE, Gaithersburg
22. Thomas GC (1996) Fire resistance of light timer framed walls and floors. University of Canterbury, Canterbury
23. Babrauskas V (2003) Ignition handbook ISBN 0-9728111-3-3, Lib. of Congr. #2003090333, Fire Science Publishers, WA, USA
24. Quintiere JG (1998) Principals of fire behavior. Delmar, Albany
25. Wickström U (2015) New formula for calculating time to ignition of semi-infinite solids, fire and materials. Published online Library (wileyonlinelibrary.com). doi: 10.1002/fam2303
26. Burley NA et al (1978) The Nicrosil versus Nisil thermocouple: properties and thermoelectric reference data. National Bureau of Standards, NBS MN-161
27. Lattimer B (2008) Heat fluxes from fires to surfaces, 4th edn, SFPE handbook of fire protection engineering. National Fire Protection Association, Quincy
28. Wickström U (1994) The plate thermometer—a simple instrument for reaching harmonized fire resistance tests. Fire Technology, Second Quarter, pp 195–208
29. Wickström U, Duthinh D, McGrattan K (2007) Adiabatic surface temperature for calculating heat transfer to fires exposed structures, Interflam 2007, London, Sep 3-5, 2007, pp 943
30. Wickström U, Jansson R, Tuovinen H (2009) Experiments and theory on heat transfer and temperature analysis of fire exposed steel beams. SP Report 2009:19, ISBN 978-91-86319-03-8
31. Wickström U, Robbins A, Baker G (2011) The use of adiabatic surface temperature to design structures for fire exposure. J Struct Fire Eng 2(1):21–28
32. Ingason H, Wickström U (2007) Measuring incident heat flux using the plate thermometer. Fire saf J 42:161–166
33. Häggkvist A, Sjöström J, Wickström U (2013) Using plate thermometer measurements for calculating incident heat radiation. J Fire Sci 31:166
34. Sjöström J et al (2015) Thermal exposure from large scale ethanol fuel pool fires. Fire Saf J 78 (2015):229–237. doi:10.1016/j.firesaf.2015.09.003
35. The European Committee for Standardisation, CEN (2005) EN 1991-1-2, Eurocode 1: design of steel structures—general rules—structural fire design. The European Committee for Standardisation, CEN, Brussels
36. Magnusson SE, Thelandersson S (1970) Temperature-time curves for the complete process of fire development- a theoretical study of wood fuels in enclosed spaces. Acta Politechnica Scandinavica, Ci 65, Stockholm
37. Wickström U (1985) Application of the standard fire curve for expressing natural fires for design purposes. In: Harmathy TZ (ed) Fire safety: science and engineering, ASTM STP 882. American Society of Testing and Materials, Philadelphia, pp 145–159
38. Babrauskas V, Williamson RB (1978) Post-flashover compartment fires: basis of a theoretical model. Fire Mater 2:39–53
39. Zukoski EE, Kubota T, Cetegen B (1980) Entrainment in fire plumes. Fire Saf J 3:107–121
40. Karlsson B, Quintiere JG (2000) Enclosure fire dynamics. CRC Press, Boca Raton
41. Evegren F, Wickström U (2015) New approach to estimate temperatures in pre-flashover fires: lumped heat case. Fire Saf J 72(2015):77–86
42. Wickström U (1985) Temperature analysis of heavily-insulated steel structures exposed to fire. Fire Saf J 5:281–285

43. Melinek SJ, Thomas PH (1987) Heat flow to insulated steel. Fire Saf J 12:1–8
44. Wang ZH, Kang HT (2006) Sensitivity study of time delay coefficient of heat transfer formulations for insulated steel members exposed to fires. Fire Saf J 41:31–38
45. Wickström U (1982) Fire Saf J 4:219, 1981
46. Wickstrom U (2005) Comments on the calculation of temperature in fire-exposed bare steel structures in prEN 1993-1-2: Eurocode 3-design of steel structures- Part 1-2: general rules-structural fire design. Fire Saf J 40:191–192
47. Jansson R (2013) Fire spalling of concrete—a historical overview. Key note at the 3rd International RILEM Workshop on Concrete Spalling due to Fire Exposure, Paris
48. Wickström U (1986) A very simple method for estimating temperature in fire exposed concrete structures. In: Grayson SJ, Smith DA (eds.), Proceedings of new technology to reduce fire losses and costs. Elsevier, New York
49. Wickström U (1985) Application of the standard fire curve for expressing natural fires for design purposes. In: Harmathy TZ (ed) Fire safety: science and engineering, ASTM STP 882 (pp 145–159). American Society of Testing and Materials, Philadelphia
50. Ingason H, Li YZ, Lönnermark A (2015) Tunnel fire dynamics. Springer, New York
51. Schneider U, Horvath J (2006) Brandschutz-Praxis in Tunnelbauten. Bauwerk Verlag GmbH, Berlin
52. Wickström U, Hadziselimovic E (1996) Equivalent concrete layer thickness of a fire protection insulation layer. Fire Saf J 26:295–302
53. Badders BL, Mehaffey JR, Richardson LR (2006) Using commercial FEA-software packages to model the fire performance of exposed glulam beams. In: Fourth International Workshop "Structures in Fire", Aveiro
54. Fung FCW (1977) A computer program for the thermal analysis of the fire endurance of construction walls, NBSIR 77.1260. National Bureau of Standards, Washington, DC
55. Gammon BW (1987) Reliability analysis of wood-frame wall assemblies exposed to fire. Ph.D. Dissertation, University of California, Berkeley
56. Bisby LA, Frangi A (2015) Special issue on timber in fire. Fire Technol 51(6):1275–1277
57. König J (2005) Structural fire design according to Eurocode 5—design rules and their background. Fire and materials 29(3):147–163
58. Östman B et al (2010) Fire safety in timber buildings. SP Report 19, SP National Research Institute of Sweden, Boras

Printed in the United States
By Bookmasters

Printed in the United States
By Bookmasters